Scrunched Cube Electron Shell and Bonding Periodic Chart of Elements

026-Fe Iron

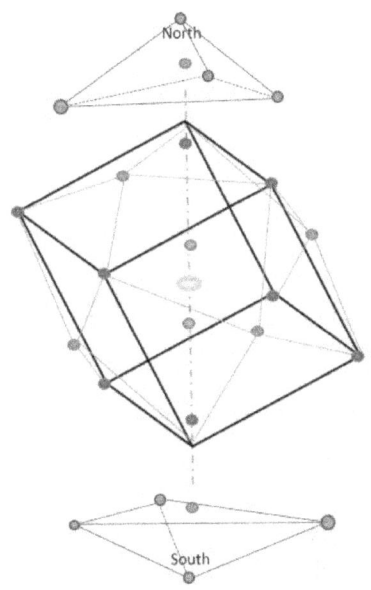

Understanding the 3D Geometry of Electron Shell Placement and Molecular Bonding Angles in the Scrunched Cube Atomic Model

By Arno Vigen

© 2016 E Arno Vigen

Simple Words to Understand . . . Chemistry, Elements, and Bonds

Why does a Nucleus Stay Together If Protons Repel?

- A Nucleus is Just . . . a Magnetic Ring

Why Don't Electrons Fall into the Opposite-Charged Nucleus?

- Electrons are Just . . . Frightened by Nucleus Magnetics

Electron Shell Chemistry Is Just . . . Scrunched Cube Geometry

- Why are electron shells in sets of 2, then 8, then 8 and such? Can we improve Pauli-aufbau?

Publication: 2017

Scrunched Cube 3D Periodic Chart of Elements Geometry Electron Placements and Bonding

- Understanding Shells for Each Periodic Chart Elements

Scrunched Cube Molecular Bonding

- Understanding Molecular Bonding in the Scrunched Cube Atomic Model

Scrunched Cube for Chemical Reactions

- Understanding Chemical Interactions in the Scrunched Cube Atomic Model

What Makes a Molecule Solid, Liquid, or Gas?

- And Why is Gas of Every Element the Same Volume (a mole)?

Simple Words to Understand . . . Gravity and Other Forces

Gravity is Just . . . That Electrons are a Little Closer

- Explaining Gravity from the basics of Electromagnetism and Explaining Why Observed Mass Changes

The Five Continuous Fundamental Electromagnetic Forces: Reconnecting Newton into the Chemistry and Particle Physics

- Resolving Strong Force, Weak Force, Bonding Force, Gravity, Mass and $E = mc^2$ via the basics of Electromagnetism as One Continuous Function

Publication: 2017

Does Time and Space Really Warp?

- Replacing Electron-Shell Radius for Time-Space Factors in formulas such as the General Theory of Relativity

How Are Electricity and Magnetism Linked?

- Exploring the Fundamental Linkage of Charge and Magnetism

Simple Words to Understand . . . Personality

Visual Astrology: Fun, Support, Security, and Growth

- Astrology 'signs' archetypes are based upon powerful traits to understand people

Visual Astrology Relationships

- What happens when 'sign' personalities interact

Visual Astrology and Jung

- Astrology 'signs' archetypes actually predict all the Jungian 4 archetypes

Simple Words to Understand . . . Communications

Decision Matrix® Writing

- Persuasion is based making arguments at the correct strength in a certain order.

GATESOUP® Writing

- **G**oal, **A**udience, **T**heme, **E**nough **E**lements, **S**upport and the rest

KADARF – Six Simple Steps to Success

Kedarf® Grammar and Composition Explained

- Defining the Parts of Speech, Paragraph Structure and More in Usable Terms

Table of Contents

Scrunched Cube Periodic Chart of Elements 7
History 9
    A Different Approach 13
    Addressing the Counter Arguments 16
        Moving Electrons Create Magnetic Fields 16
        The Prior Art Calculation with Angular Momentum Works So Well 17
    Arno Vigen Scrunched Cube (AVSC) Model 18
Full Shells 24
Subshells 27
Impact of Magnetic Field = Scrunch 32
Exception #1 – Equatorial, Transitional Subshells 33
Exception #2 – Occasional Mismatched Hemispheres 35
Impact on the Periodic Chart 36
    Models Similar at Far Left and Far Right 37
    Major Difference is the Structure Changes at +8 in any Subshell 38
Shell Description of Elements 51
    Full Shells Give Bonding Positions When Not Full 51
    Nucleus, Its Magnetic Field, Electrons, and Bonding Positions 52
    Polar View Based from One Magnetic Pole 52
    Equatorial View Based Upon the Magnetic Poles 53
    Structure View Based Upon the Magnetic Poles 58

Electrons Shell Configuration and Bonding Locations of Chart of Elements 71

Angles by 3D Geometry 180

AVSC and Quantum Mechanics 190

    Shell Number 191

    Subshell Angle and Structure 191

    External Magnetic Quantum Number 193

    Spin Quantum Number 194

Endnotes 195

Scrunched Cube Periodic Chart of Elements

Understanding chemistry can be easy using 3D geometry. It builds based upon the fundamental magnetic field of the nucleus of each atom. Electrons shells build first at the poles; that is only 2. Next the lowest energy is 4-point tetrahedron anchored at two poles, so 4 x 2 = 8. It keeps building, and the properties of each element related to that 3D geometry of how to best fit together layers (shells) given existing shells:

- How many electrons, repelling each other, can fit in a 2-magnetic-pole anchored sphere for the full shells. When full, another shell, larger, then starts.

- For subshells, what is the 3D structure based upon the electron repulsions for that layer, and the layers below. How do they build sphere upon sphere in each hemisphere?

The structures are consistent whenever the situation includes the same number of electrons, protons and neutrons. The number of atoms is called the element.

This 3D structure makes all the properties of elements: electron spacing, electron angles, bonding angles, positions where electrons exchange which creates discrete wavelengths which come is sets that define electromagnetic spectrum, ability to hold magnetism, and the release energy for electrons, and strength of bonding, and melting point and bonding point. You get it – lots depends on the understanding of the 3D geometry of electrons shells and subshells.

This book will explore the structure details for each element in the periodic chart.

Big hugs. Let's get going.

History

Ernest Rutherford in the early 1900's determined by documented experiment that an atom consisted of a nucleus and many electrons in a field surrounding it. He proposed that electrons orbit, in the planetary-like model. That is, the charge force (electrostatic) charge pulls them together, and angular momentum keeps them apart. It is like gravity.

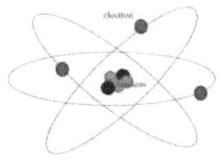

Atomic Planetary Model

In 1913, Niels Bohr updated this by applying the 'quantum of energy' determined by Max Planck based in part on prior work of Ludwig Boltzmann, and theorized more generally by Albert Einstein. Bohr came to a specific calculation of the energy levels, the 'quanta' that are the jumps for 001-H Hydrogen electron energy levels when overloaded (ionized) with many electrons. This described spectrum lines well.

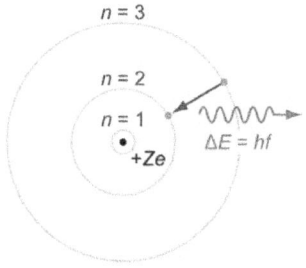

The Bohr calculations worked perfectly for 001-H Hydrogen, but would not match up for other elements. To this day, these calculations work for 001-H Hydrogen, but an improved formula, even in quantum mechanics, do not determine a) bonding angles, and b) electron shell structure.

However, it was entirely based upon Rutherford's angular momentum. That is the outward pressure of the speed equals the inward pull of the nucleus proton-electron attraction.

That is, a nucleus (101) has an electron (102) in an orbit (103). The speed of that electron creates momentum (104) which in the direction of the nucleus has a project force (105) in that direction/dimension that equals the proton-electron electrostatic charge attraction (106).

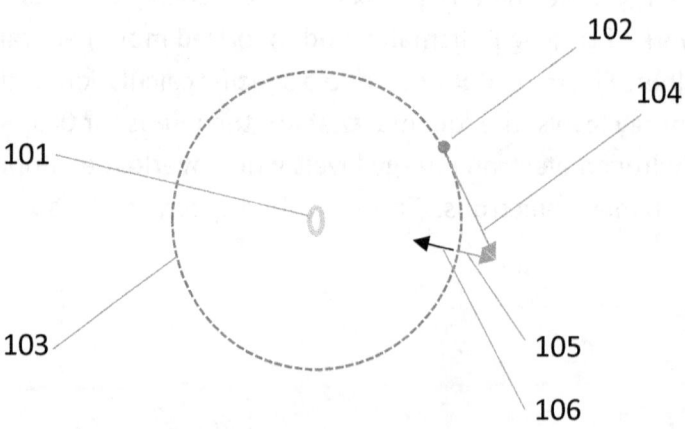

This physical model was the grounds for the calculations, as Sommerfeld, Pauli, Fermi and Schrodinger used the

Rutherford angular momentum model with the Bohr quantum jumps model and applied matrix calculation to determine the energy levels and spectrum for elements. Then, Swinger and Linus Pauling fixed the angular momentum calculation even more to get electrons energy, spectrum, and such for 001-H predicted to 10+ decimals.

This line of theory progressed to Schrodinger's Equation which uses four quantum numbers to create an excellent prediction of energy levels, spectrum, and some other properties of the elements of the periodic table. It is a superb system based upon understanding the harmonics of electrons in certain groupings (the shells).

However, all that quantum mechanics does not predict the periodic chart. Specifically, the bonding angles and reasoning for why the harmonic sets at 2, 8, 8, 18, 18, 32, 32, 50, and 50. There is nothing to say that the system should not be 2, 4, 8, 16, 32, 64 as the progression. Schrodinger's equation works equally as well upon that progression.

All the formulas work. The math is superb, but it has weaknesses. It moves off physical to only statistical models where the physical forces are a) non-continuous, and b) not connected to a physical model.

In fact, the story is that Schrodinger and Einstein got together one weekend, and discussed the struggle that the quantum model veered into only statistical solutions. Soon after that Schrodinger wrote about that his cat could exist and not-exist at the same time. Most people take Schrodinger at his word, and seriously think that reality is existence and non-existence at the same time; whereas, I think that Einstein and he, Einstein especially, were showing the fallacy of the entire science community taking on such a position. Of course, the funny thing is that those two were the ones that started and perfected that existential system.

I prefer A is A. There is a physical force driving these structures, and a knowable structure for each of the results in the Periodic Table. That is the shells and subshells are based upon:

A Scrunched Cube

Well, just the 2nd and 3rd shells are the Scrunched Cube. Yet, the rest of the atomic structure, the periodic chart of elements, builds on that that core.

A Different Approach

In his 1913 paper, Bohr stated that electron remain in "stationary orbits"[i] to describe the combination of quantum levels with angular momentum. However, Bohr said one thing, but actually failed to recognize the juxtaposition of that statement. "Stationary" and "orbits" by definition cannot mix. A stationary particle cannot orbit; it falls into the central object pulling it.

That is, it cannot without a trick.

Instead, the periodic chart of the Arno Vigen Scrunched Cube (AVSC) model does something very different. The core, different assumption of AVSC is that electrons rotate as a group.

Electrons rotate as a group

In fact, this as a group has actually been utilized, but not understood as a physical phenomenon. The basic method of the Schrodinger equation is to solve a matrix algebra problem. That is solving multiple equations at the same time 'as a group' with the same elements. That is,

$x + y = 2$

$x - y = 0$

This becomes in fancy math an (x,y) matrix.

| 1 | 1 | | 2 |

$$1 \quad -1 \quad = \quad 0$$

So, many of the quantum mechanics bring solutions to electrons in sets. Those different equations are different particles that must solve with the same forces – as a group.

For course, to rotate as a group, there must be a force that makes that fitting together process work. Otherwise, electrons would not have subshells and such.

An Electron Shell Structure Fills Based Upon the Nucleus Magnetic Field Repulsion versus the Electrons

Notice that a magnetic field has the funny shape with more repulsion at the sides (towards the equator), than at the magnetics poles (Yes, nucleus-electron magnetic repulsion and strengths are exact opposites of normal thinking that magnetic attract at the poles).

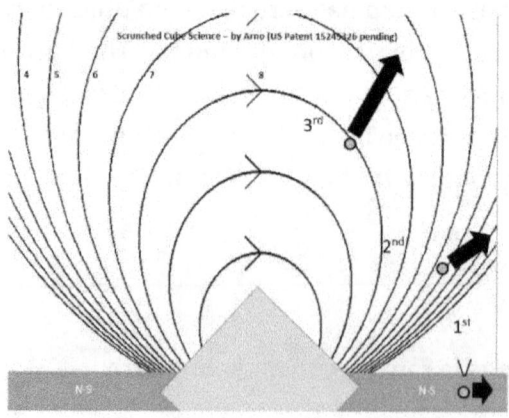

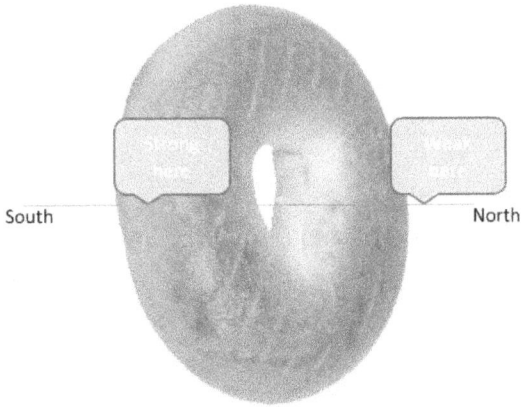

That shape is different than the perfect spherical one of charge (electrostatic). That mix of push and pull from two different shaped forces creates a differentiated model that drives electrons into the different slots of the periodic chart.

It is that push and pull that put electrons into specific slots for each element. It is that push and pull that makes the positions, angles, energy-levels, distances, spectrum unique for each element, and even for each isotope of an element.

Addressing the Counter Arguments

Moving Electrons Create Magnetic Fields

Yes, the combination of magnetic field and electrons has been documented to create radiation. In my telecommunications industry, we all know that moving electrons in a wire create a magnetic field around that wire. Think Ampere's Law and Maxwell's equations.

However, the combination of electrons rotating in the magnetic field that is also rotating changes the framework in question. That combination is the trick. From the perspective of the nucleus, the electrons are not generally moving. Think relativity. In every atomic system, the electrons are stable in their system. Our thinking that those electrons move, that is, orbit, from our perspective, outside the system, is exactly the sort of mistakes that Einstein recognized, and against which he admonished.

Any internal movements of the electrons settling into the AVSC 3D geometric locations are small enough from that nucleus perspective as not to trigger a quantum jump normally. From there, actual events, like bonding or ionization, can trigger the observed radiation, quantum jumps, when those other events do occur.

This system does not break those know observations.

The Prior Art Calculation with Angular Momentum Works So Well

Given the success from Bohr through Schrodinger and continuing to Swinger and Feynman, how can a different formula exist?

The basic structure from Bohr to quantum mechanics is that adding momentum (mv) will offset the 1/distance-square of charge. Yet, the 'v' is velocity which is distance/time. Therefore, the solution adds a distance-factor which can be the same as moving from 1/distance-squared to 1/distance-cubed. Both angular momentum and magnetic change the equation by distance-1st power.

Isolating momentum changes distance by the first power.

$$mv = \frac{mx}{time} r$$

Using magnetics, also changes distance by the first power.

$$\frac{kqq}{r^2} = \frac{M}{r^3} ::\gg \frac{kqq}{M} r$$

This is not a rigorous proof, but it gets the sense of a path where magnetics can replace most of the angular momentum. Therefore, the successes of quantum mechanics did not utilize the momentum in the more advanced calculations. Quantum mechanics really works in whole numbers of quantum jumps of linked sets as a statistical group to understand the harmonics of the combined system of electrons. Those results would be the same with either formula.

Arno Vigen Scrunched Cube (AVSC) Model

The AVSC model adds a factor before angular momentum. "It takes a fork in the road between Rutherford and Bohr, but eventually has a path back to Pauli, Fermi, and Schrodinger with a fuller, better set of attributes that complete the chemistry picture. It both solves bonding angles, and it feeds the factors of the quantum numbers needed for quantum mechanics."

In fact, that nucleus-magnetic factor applies then makes the Bohr-used angular momentum force a secondary factor that applies in the normal course along with other factors – such as whole atom speed, external magnetic forces, speed relative to the speed of light, and other subatomic particles entering the system. The new factor is larger than angular momentum at those distances.

The Nucleus Magnetic Field Repulsion versus the Charge Attraction Creates the Reason, the Net-Force that Keeps Electrons Away from the Nucleus

The new force is a nucleus-electron, magnetic-shaped repulsion force based upon the number of particles (protons + neutrons = 'atomic mass') and the angle of the particles relative to the magnetic axis of each other.

It is that bagel shape of a magnetic field.

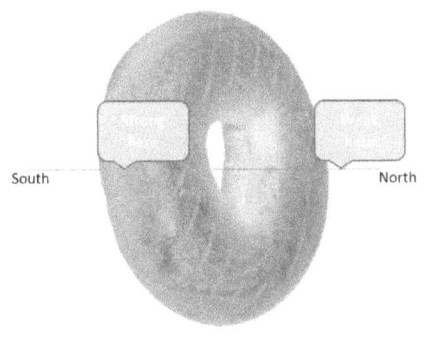

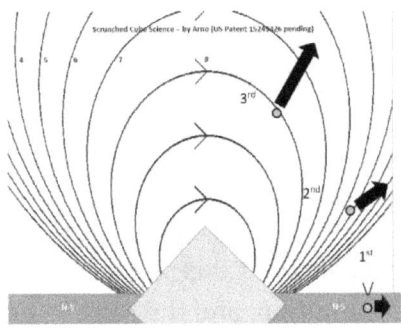

That replaces the prior art with the AVSC model of nucleus-electron forces.

A nucleus (201) has an electron (202) in an orbit (203). The main, largest force is the proton-electron electrostatic charge attraction (206) which pulls the electron towards the nucleus and maintains the same strength in any directions (spherical). The major, largest offset is a nucleus-electron magnetic-repulsion force (207) based upon the distances and an angle versus the magnetic axis of the nucleus (208). That still leaves a smaller force vector which is the portion of the orbit speed of that electron creates momentum (204) which in the direction of the nucleus has a project force (205) perpendicular to the orbit.

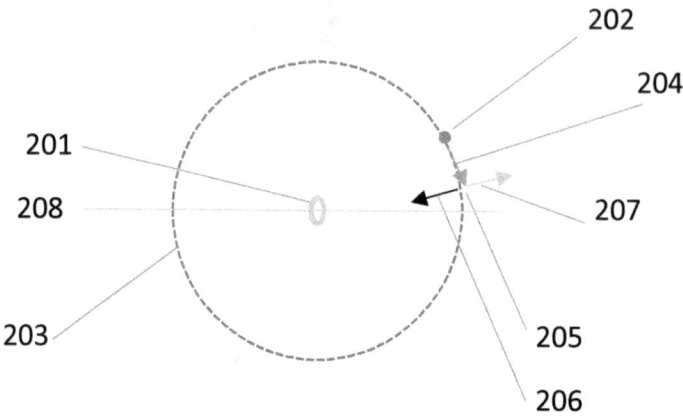

There are other factors, but this picture describes the new factor, magnetic repulsion, and the relative decrease in angular momentum of the prior art. Angular momentum (205) is tiny now compared to the other two (206, 207).

The formula (without other forces for simplicity) is the combination of the charge force (Coulomb's Law) with magnetic force using a simplified formula using a similar factors as found in Biot-Savant (although other magnetic-force integrals also work, but are more complex):

$$F = \frac{kq_0 q_1}{d^2} + \frac{M(n)\sqrt{1 + 3\cos(\theta_{n,e})} * M(e)\sqrt{1 + 3\cos(\theta_{n,e})}}{d^3}$$

k = Coulomb's constant

q = sum of charge particles in each element.

Please note that since the protons are + and the electrons are '-' that the $q_0 q_1$ is negative (attractive) in every case.

d = distance between the particles

M(x) = the gross magnetic force of that set of particles

$\theta_{x,y}$ = the angle to 'y' versus the magnetic field of 'x'. That is, the angle is calculated based a vertex of the nucleus with two vectors of 1) particle 'x' to an external point on its own magnetic pole, and 2) particle 'x' to 'y'

Specifically, the Biot-Savant has a '+T' inside the square root for a tiny factor not address properly in the formula which applies at distances in the small end of the range, but that work in the full integral. A full expression as a surface integral would look like:

$$F = \frac{k q_0 q_1}{d^2} + \frac{M(n) \iint S[B(n, \theta_{n,e})] * M(e) \iint S[B(n, \theta_{e,n})]}{d^3}$$

The calculation is further complicated because the nucleus particles are not a straight line of magnetic. Therefore, they build into their own structure which makes the actual calculation of $B(n, \theta_{n,e})$ another level of complicity.

All that is not addressed here. To create a periodic chart, the simplified structure is various shells build of subshells from on magnetic pole to a slightly bigger radius at the equator back to the 'spin' opposite one electron at the opposite pole;

it does this shell by shell slowly adding particles as there is more room in each hemisphere. It is two (2x) hemispheres by the ROUNDUP[(n+1)/2,0) which is the traditional 2, 8, 8, 18, 18, 32, 32, 50, and 50 already well know.

The subshells structure builds in rings at specific angles based upon the energy a) of the nucleus magnetic field, and b) the layers below. They go from one pole to the other as:

1/1

1/3/3/1

1/3/5/5/3/1

1/3/5/7/7/5/3/1

The subshells are all anchors by the two electrons at the low energy magnetic poles. The nucleus distance (r=radius) goes from smaller at the axis to larger towards the equator.

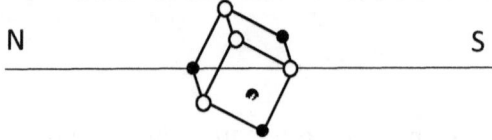

The AVSC describes the full shell structures as:

Magnetic Poles of Axis

Scrunched Cube

Upper Scrunched Sphere

<u>V</u>ery Big Scrunched Sphere

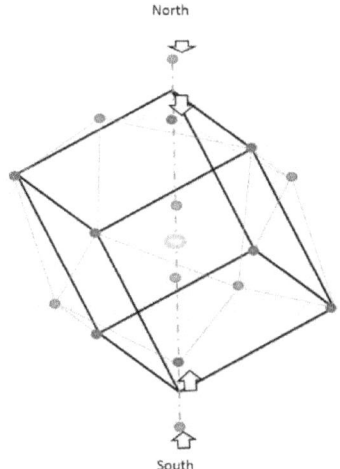

Finally, the even shells can fit between the electrons on the odd shells to create two layers at each size limit. For subshell 2. That is why these structure goes 8, then 8, and 32, then 32 and such.

This summary was intense, so in the next section, I will describe each of those elements more slowly and with lots of pictures.

Full Shells

The current periodic table uses a building structure based upon layering shells of size 2, 8, 8, 18, 18, 32, 32 in order. That part is well understood. From there, the subshell rule of the last half century assuming that further subshells 2,6,10,14 that build in a particular filling order called the German word for filling – 'aufbau'. Yet, this 'aufbau' current subshell logic has a process of fillings the 2 at the left of that listing then builds from the right. It is a fixed pattern different than AVSC.

       Shell filling rule: 2,14,10,6

These are typically labeled:

       Shell filling rule: s,f,d,p

But that naming actually floats back to naming of lower shells, so the 'aufbau' order for say 085-At becomes:

       Shell filling rule: $6s2, 4f14, 5d10, 6p5$

The f-shell (which maxes as 14 electrons) seems to be at two layers below, but is not filled until after 6s2 and does not occur until the 6th shell.

While it works in most cases, this pattern is not matched with a physical model with many known exceptions (See 021-Sc Scandium exception and others).

The Arno Vigen Scrunched Cube (AVSC) model considers shell filling in a few basic rules (with certain transitional structures discussed later):

- Electrons build with the magnetic field from the nucleus particles. That makes for two (2) hemispheres build in opposing pairs. That is the physical reasoning for the 'spin' limit of two (2) electrons of the Pauli exclusion rule. For our purposes, this resolves what we call the 'z' polar-axis dimension in an (x,y,z) coordinates or the polar axis for measuring the '<i' inclination (the latitude) dimension in spherical (r,<i,<j) coordinate system (where <i=0).

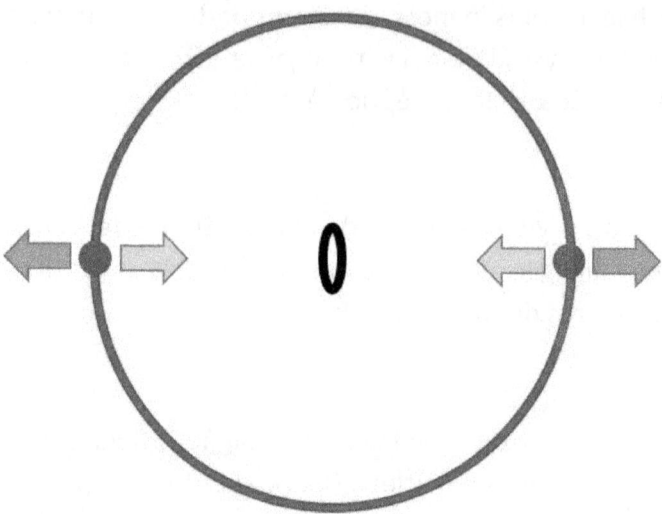

This resolves one dimension in each shell/subshell. It is limited to 1 or 2 electrons – each at the exact opposite position in a sphere.

- The subshells build from each end towards the center, generally filling the ends first which are lower magnetic-like repulsion.
  Each shell gets bigger because it must build on the circle of inner electrons. That makes the full shells 1-squared, 2-squared, 3-squared.
  The next layer includes one from the lower level, plus one extra row in the x, and one extra row in the y direction. As it builds on the inner layer, more electrons can fit into that next layer.

1

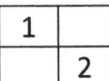

Subshells

The steps to create squares basically build subshells in steps. Let's start with building squares

1	=	1	=	1-squared
1 + 3	=	4	=	2-squared

1 + 3 + 5	=	9	=	3-squared
1 + 3 + 5 + 7	=	16	=	4-squared
Above + 9	=	25	=	5-squared

To make a square you take a base, and add one in each dimension (+2) in that hemisphere, and put one on top of all the existing ones.

Think about the first shell. It has one electron (1m).

Then the second shell can have 1 (2m-shell) on top (towards the magnetic pole, but also needs the 3 to surround the rest of the one (2c-shells).

In this case, the new shell completely surrounds that 1m shell of electron; the inner shell is hidden in the middle. That inner shell is what pushes the central electron up (outwards towards the magnetic pole).

Then the second shell can have 1 (4m-shell) to top, but also needs the 3 to surround the rest of the one (4t atop 3f-shells in one direction, and around the 3m-1 electron), then it needs another ring of 5 around the 3f-three electrons.

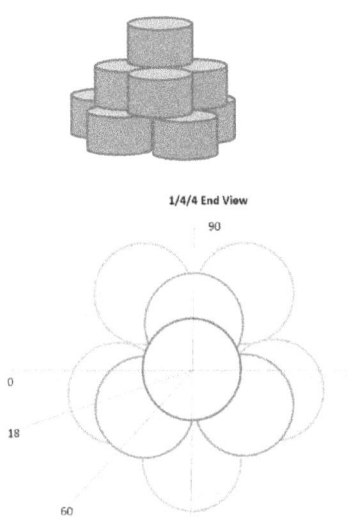

Again, that structure completely encases the 1/3 structure inside it, with the inner structure pushing up the 3 layer once, and the 1 layer twice.

- These build in from the low-energy towards the equator. That makes the full shells:

<pre>
              Eq.
     N   S
       1/3/5/7/7/5/3/1
</pre>

Here is a graphic of the shells in one hemisphere building up.

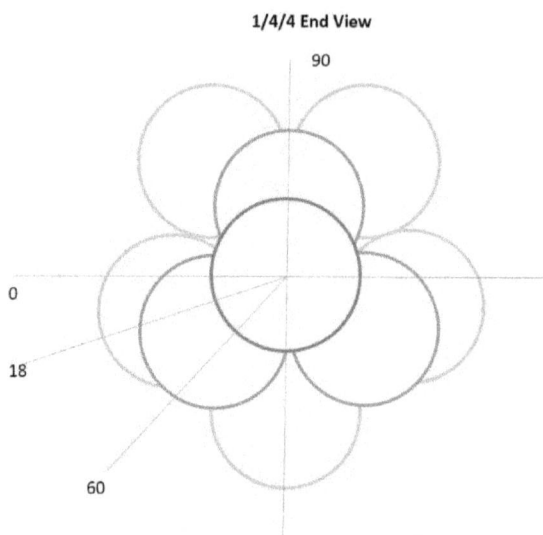

Notice that to create the subshell counts that increase by 2 in each step. That is the 1-x and 1-y of the add-on layer. 1>3, 3>5, 5>7. That +2 occurs because you build in 2-dimensions.

As you can calculate, one type of full shell has 16 (4-squared) in each hemisphere. That makes 32 which experiments support is the size of the full outer shell for the noble gas 036-Kr Krypton or 054-Xe Xenon.

When less than full, those electrons actually fill positions from the axis so that 010-Ar Argon would be only the two outside subshells filled in each hemisphere:

N　　　　　　S

1/3/n-a/n-a/n-a/n-a/3/1

And if next subshell has only '1' for 011-Na Sodium, then only one side is filled. (I am not sure if it is north-first or south-first, so I will choose north for convenience):

N　　　　　　S

1/0/0/0/0/0/0/0

Or, that 085-At describe earlier would be:

N　　　　　　S

1/3/5/7/6/5/3/1

Before the full 086-Rn Radon of:

N　　　　　　S

1/3/5/7/7/5/3/1

Impact of Magnetic Field = Scrunch

Because the repulsion is less at the magnetic poles, this filling rule makes all the know results for subshell energy levels still apply. The magnetic poles are closer, and the others have a larger radius to the nucleus a) moving towards the equator, and b) based upon fitting above inner layers.

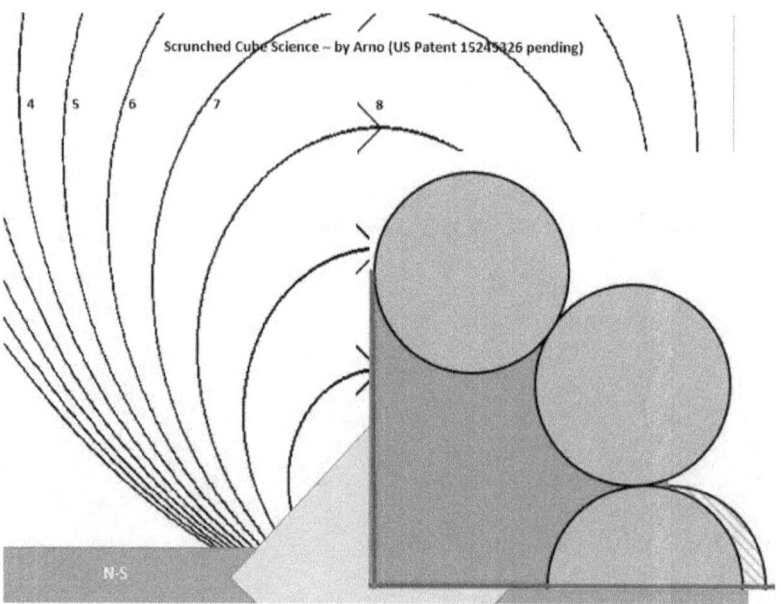

The prior art finds this in the low-energy observations of 2s versus 2p, or 3s, versus 2p, or even 5s versus 4d. The added electrons get added last toward the higher energy equator relative to the nucleus magnetic field.

Exception #1 – Equatorial, Transitional Subshells

When there are 1, 2, or 3 electrons only in new subshells, sometimes the math works that they can slip to the equator instead of the angle of the full shell structure.

However, once the subshell reaches 4, then 90 push of four electrons at the equator makes all four flop into the full shell alignment. That is, the equatorial subshells are transitional only.

See that 3 electrons are at 120 degrees so the other two are all the way on the other side and so don't repulse.

Two Repulsion Forces

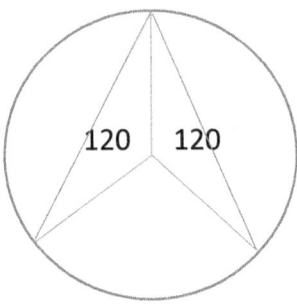

The distance is the square-root of (1.5-squared+cos(60)-squared)

However, repulsion is 1/squared, so the force is >1/3 x two particles = <2/3.

But, at 4 electrons, they sit at 90 degrees at the same longitude so the pressure is 1/-distance-squared which is ½ for each. But there are two, so the pressure is a full 1 pressure just from the 2 electrons at 90 degrees. Most of the time, the full structure north-and-south then has less energy.

Three Repulsion Forces

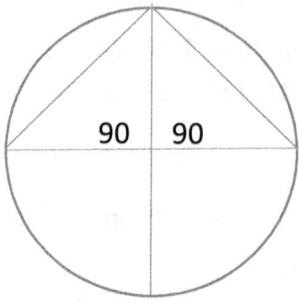

The two side particles' distance is the square-root of (1-squared+1-squared).

However, repulsion is 1/squared, so the force is >1/2 x two particles = 1 plus the opposite ½, so 1-1/2 which is more than double when only 3 particles.

Equator Repulsion from each other 3 particles = 0.7 F

Equator Repulsion from each other 3 particles = 1.5 F

Exception #2 – Occasional Mismatched Hemispheres

At 006-C Carbon, there is no lower layers, so the electrons best repulse against each other. That means that the four (4) Shell-2 electrons can act like just four electrons and form the smaller structure possible in 3D, a 4-corner tetrahedron.

Of course, one electron tends to find the magnetic pole low energy, that makes the 006-C an unusual [2m1,2c3] configuration.

The unusual part is that e-2c3 are all in one hemisphere.

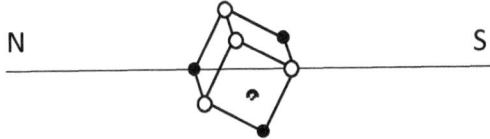

A similar challenge occurs with 007-N Nitrogen at 2m1,2c3, again all e-2c3 electrons sit in one hemisphere.

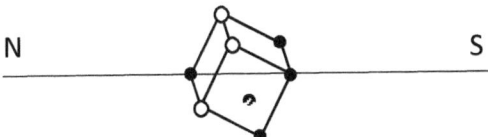

Impact on the Periodic Chart

In general, the resulting periodic chart of elements in the prior art versus in the AVSC model, have many similarities.

In periodic chart, the shells build, after 2, right to left – even though the size obviously builds left to right. Therefore, the periodic chart has six locked columns at the left, yet those the properties do not flow well down the columns.

Maybe a picture of the properties of metalloids per a current textbook will show this discrepancy. The metalloids are black boxes, but they do not fall in a column:

P-Shell			1	2	3	4	5	6
01-H								02-He
03-Li	04-Be			06-C	07-N	08-O	09-F	10-Ne
11-Na			13-Al		15-P		17-Cl	18-Ar
19-K			31-Ga			34-Se		36-Kr
				50-Sn			53-I	54-Xe
					83-Bi		85-At	86-Rn

The prior art model assumes that properties build:

    2   p-subshell for every shell

    14   e-subshell for Shells 6 and up

    10   d-subshell for Shells 4 and up

    6   s-subshell for Shells 2 and up

The basic postulate for this Scrunched-Cube electron shell model and periodic chart is that the shells and subshells do not build in that manner – strictly 2, then right to left. The evidence is that like properties do not go down a column.

The Arno Scrunched Cube Atomic Model would group elements differently.

The current model orients more toward right filling, but the Arno Scrunched Cube model orients from both the left and the right in stages. It builds with the best structure only at the number of electrons.

Models Similar at Far Left and Far Right

The properties of Full and Full-Minus-1 columns are very similar. An inert gas is the same. Both models get the same results.

1	2	3-8	8+1	8+2	8+3		Full-Minus-1 Halogens	Full
01-H								02-He
03-Li	04-Be						09-F	10-Ne
11-Na			13-Al				17-Cl	18-Ar
19-K			31-Ga				35-Br	36-Kr
37-Rb	38-Sr						53-I	54-Xe
55-Cs	56-Ba						85-At	86-Rn
87-Fr	88-Ra						Halogens	Inerts

The properties are the same at the far left. Of course, that is because the aufbau model snuck in the p-shell with two electrons because the physical experiments demanded it.

1	2	3-8	8+1	8+2	8+3		Full-Minus-1 Halogens	Full
01-H								02-He
03-Li	04-Be						09-F	10-Ne
11-Na			13-Al				17-Cl	18-Ar
19-K			31-Ga				35-Br	36-Kr
37-Rb	38-Sr						53-I	54-Xe
55-Cs	56-Ba						85-At	86-Rn
87-Fr	88-Ra							
Alkali Metals	Alkali Earth Metals							

The edges are solved long before Scrunched Cube. The Scrunched Cube model is necessary to solve the complex section in the middle. What are the structures as the electrons jostle for position as the shells and subshells grow?

Major Difference is the Structure Changes at +8 in any Subshell[1]

The spectrum number of lines goes from a huge number to a few always at the 8 from the left (or 18 for Shell 6 or 7). The properties working then from the left also stay well with a left oriented set of columns.

The change is dramatic . . . and the element in columns up to that are very consistent in properties. The ones after, to the right, are also consistent, yet different than the properties of

---

[1] For Shell 6 the conversion point is 8 + 14 because the outer sphere has four (4) layers, and the extra one works like the 8-subshell of all the lower shells. This creates the electromagnetic shift at 22, but who really does chemical reactions with 065-Tb Terbium, so you can forget this fact.

the element on the left of this break point. The break in spectrum is at 8 from the left, not 10 . . .

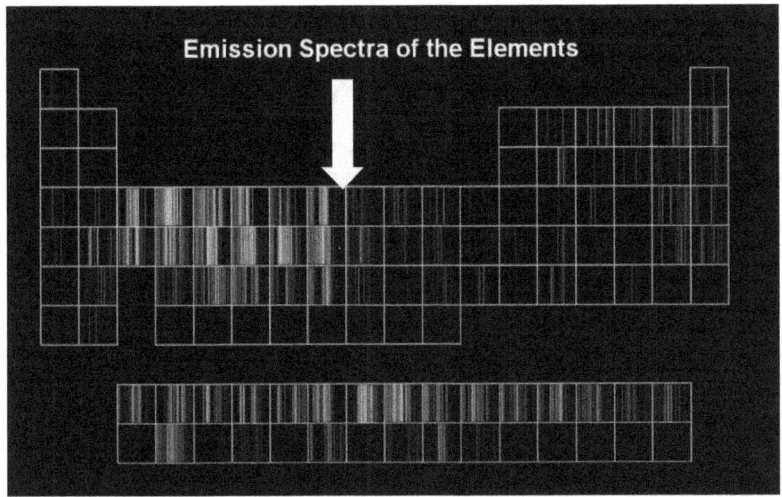

Therefore, both ends have common columns, leaving the middle to have better grouping. Therefore, the Lanthanides and Actinides are broken up and re-built with the left focus elements.

Therefore, the metalloids in the middle need a grouping that is not right-column that fits an updated physical model. The right focus does not display the physical reason why those metalloids have similar properties.

The mystery of this +8 focus is driven by the below picture of the scrunched cube model for 026-Fe Iron. This is the most stable magnetic metal. It is also have the most wavelengths for electromagnetism.

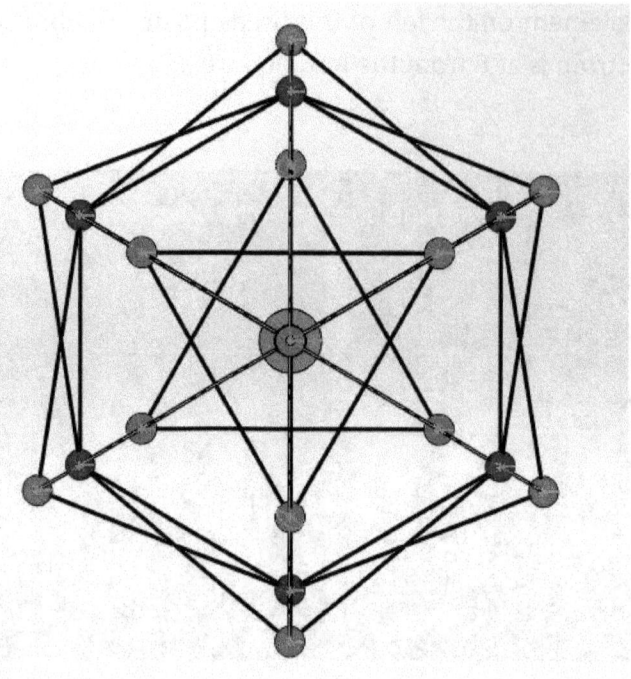

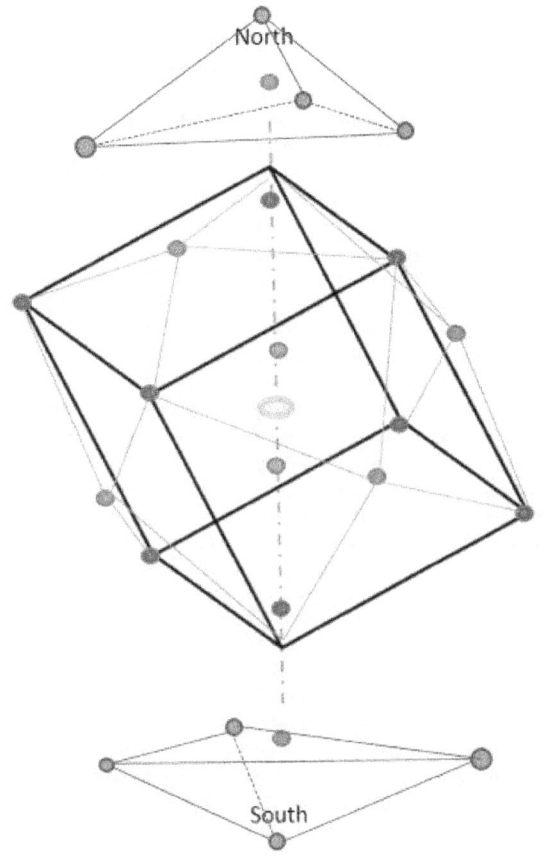

This picture is the core of the 3D geometry in the Scrunched Cube model. It describes the basic features that drive Elements to have their properties:

1) Electron shells all work based upon a magnetic field from the nucleus. That makes every location come in pairs (the Pauli exclusion explained); electrons tend to pair with one in the opposite magnetic hemisphere.

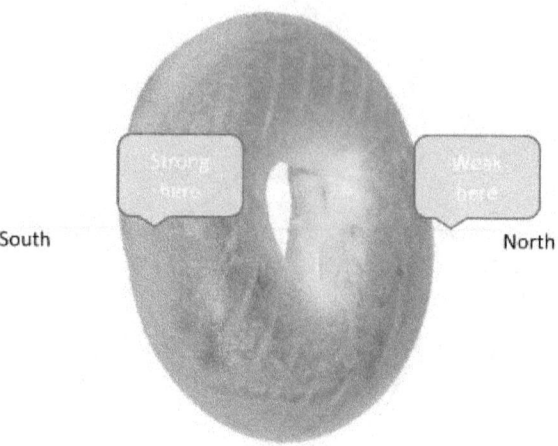

2) Shell-1 is just two at the poles. Two is all that fit.

3) Shell-2 builds the next stable structure which is a 4-point tetrahedron. Yet, because of the poles, there are two interweaved 4-points, so that is 4 x 2 = 8 which is a cube. (Replaces Pauli-aufbau.)

4) The electrons at magnetic north and south are scrunched because the magnetic field is weaker at the poles versus towards the equator. Again, a picture might help.

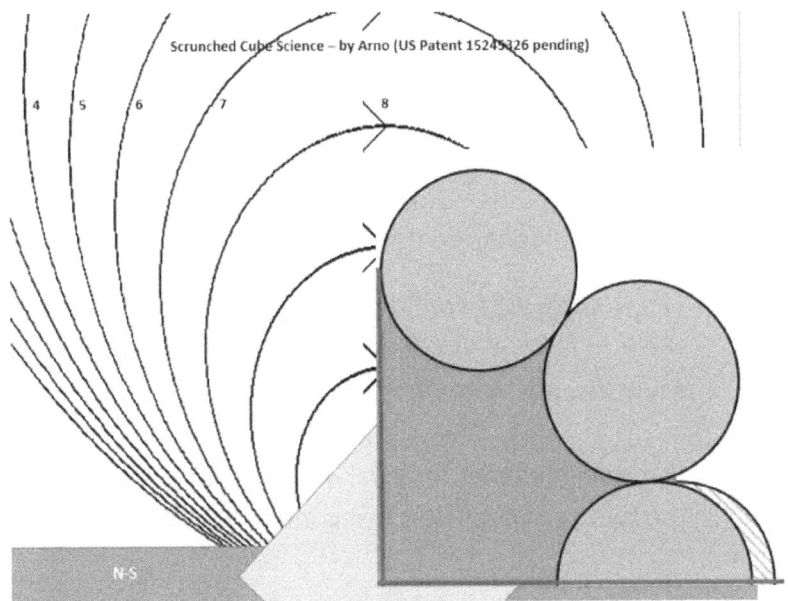

This explains why there are subshells. Electrons in subshells are not at the same distance or energy. They are different based upon the magnetic field orientation. In the old naming, this was 2s2 and 2p6 without the 3D structure. To make these understandable, I use new names for subshells: 2m2 for magnetic and 2c6 for the rest of the cube.

5) Shell-3 sits in the faces of the cube. It fills the holes.

6) The Larger Shells build from the magnetic ends. Therefore, you get eight (8) electrons in an endcap tetrahedron (resolve the transition metal aufbau failures)

That is everything in 3D in one picture. Understanding this picture and the reasons for all the chemistry properties will follow.

Reviewing before we tackle the middle sections:

- The Noble Gases would not change as those are full shells in all models. That is the far right. The full shells are not very different in the two models.

- The Halogens that are near-full also are matching in the two systems. That is second to the right.

- The Metal get separated by the endcap properties at +1 for larger shells.

- The Alkali Metals and Alkaline Earth Metals from Shell 3 up would remain. It is tough to change +1 and +2 as those are the only electrons in the extra shell.

Now, everything in 3D has intermediate structures.

Some elements have transitional shells that go away. There is math that the equator works for certain number of electrons. However, the equator never works for any full shell. It is an interesting fact that will apply when you get deep into Chemistry.

The 2nd shell is treated entirely differently and separately as the properties are not column wise. Columns only work when the larger element has a lower sphere of electrons so it can 'fit in' and thereby be 'like' the layer below. However, the 2nd Shell first three elements are more similar and should get grouped widthwise, not column wise. By the way, they are equatorial!

Among the most significant changes to the middle are:

- The transition metals would end at t6, not the traditional d10. The change of properties is significant.

- The first six Lanthanides and Actinides would fall in the Endcap group. The rest are in a large, 3-layer endcap.

- The even versus old shells have a pattern. The evens have a layer below that has different numbers of electrons, so evens have diverse properties. The odd shells tend to have more column similarities because the odd shells just fill in exactly between the even layer below so properties carry down better.

The Chart of Elements group like:

Full and Near Full Outer Shell:	
Nobel Gases	Build from the right
Halogens	Build from the right
Just Building Outer Shell:	
Alkali Metals	Build from the left
Alkali Earth Metals	Build from the left
Endcap Magnetic Metals	Build from the left
High-Melt Metals	Build from the left
Non-Multiple Shell Structures:	
Hydrogen	Small Molecules
Equatorial-Reactive	Small Molecules
Scrunched-Cube-Reactive	Small Molecules
Mid-Range Structures:	
Electrically-Most-Active Metals	Longitudinal Electrons
Poor Metals	Semi- Malleable
Metalloids	Malleable

1	1	6	2			varies	varies	1	1
001-H									02-He
Equatorial-Reactive					Scrunched-Cube-Reactive			009-Fl	10-Ne
11-Na	12-Mg			1 x	1 x		2 x	019-Cl	18-Ar
19-K	20-Ca	6 x	2 x	3 x	2 x		1 x	35-Br	36-Kr
37-Rb	38-Sr	6 x	2 x	4 x	2 x			54-I	54-Xe
55-Cs	56-Ba	20 x	2 x	5 x	1x			85-At	86-Rn
87-Fr	88-Ra								
Alkali Metal	Alkali Earth Metal	Endcap Metal	High Melt Metal	Electric Metal	Metalloid			Halogen	Inert

Notice that I split the large 'metal' group into a) Endcap Metal, b) High Melt Metal, and c) Electric Metal which have an outer shell near the equator, and so easily sent in electricity.

As such, the traditional d10 (or 14f subshell) falls into different buckets. The activities of elements in the new blocks act quite differently.

		6 x	2 x	2 x				
		6 x	2 x	2 x				
		16 x	2 x	2 x				
		16 x	2 x					

Most large, non-full structures have an endcap. That means that all the middle layer, and the endcap all have free access for electron activity (like electromagnetic spectrum, electron exchange, electron contribution in bonds or alloys).

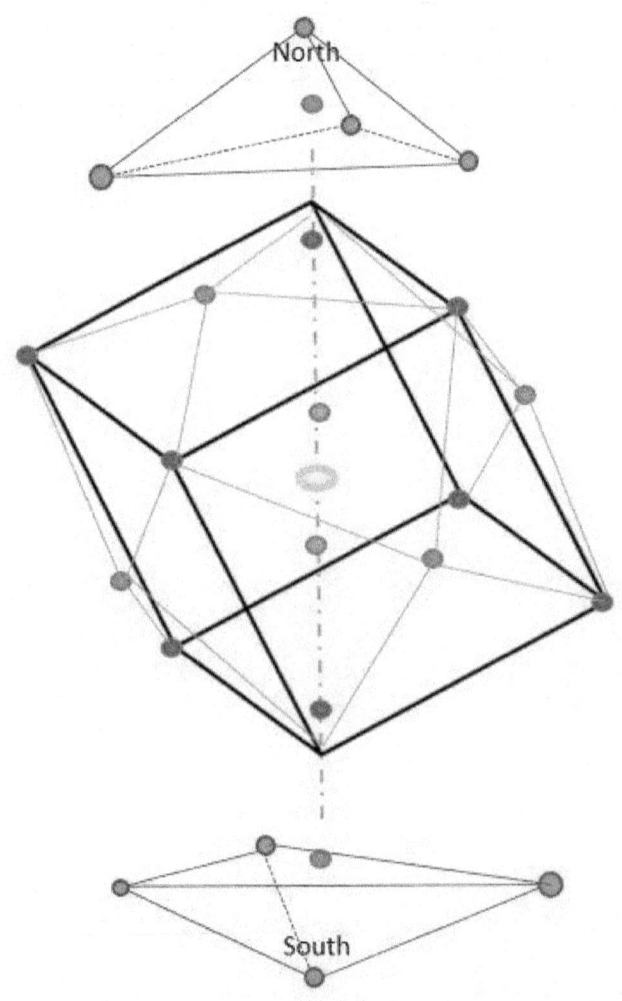

However, there is a huge change in the properties once that endcap is full and a further electron is added to the system. This occurs beyond six (m2 + t6) in the outer layer of Shell 4 and 5 and beyond eighteen (m2 + t6 + u10) in the outer layer of shell 6 and 7.

##FIX – only 3 in 4t subshell.

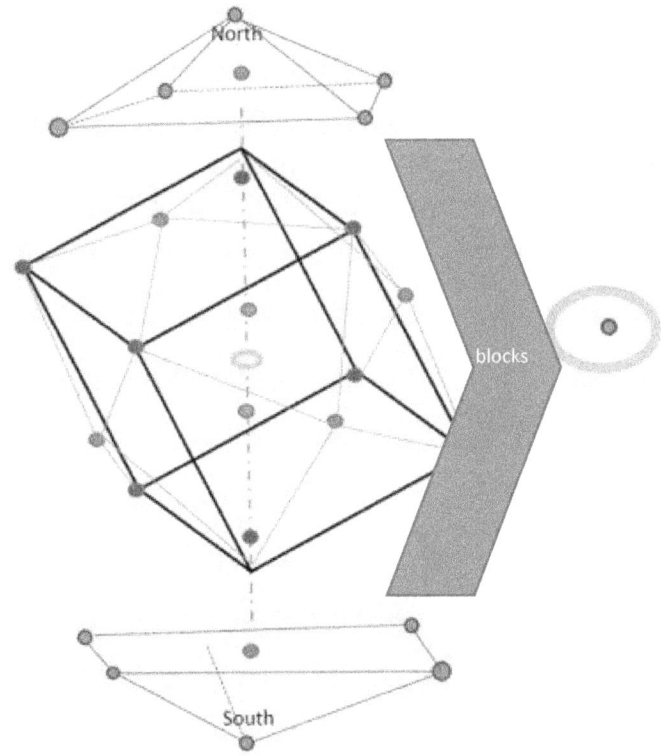

That extra electron become very dominant in electricity, in bonding/alloys interaction, in electromagnetic spectrum. The a) release energy and b) directions of release are vastly greater the u-subshell in Layer-4 or Layer-5 or v-subshell in Layer-6 or Layer-7. As such, that electron overwhelms what was a full spectrum of generally equal opportunities for the prior exposed inner layer to release electromagnetic. In 029-Cu Copper, the extra electron does the action in the clear majority of times. 029-Cu Copper e-4e3 subshell has more than 210 degrees to receive and emit radiation. This compares to about 22-1/2 degrees available for four different subshells in a 026-Fe Iron.

More importantly, the u-shell electrons are a) at the strongest repulsion location in the magnetic field and b) at a high angle to the shell structure. This is fantastic for electrical interactions.

Shell Description of Elements

Full Shells Give Bonding Positions When Not Full

In a full shell, there is a physical structure that is easiest to understand. All the positions are full with equal strength particles, the electrons. The Scrunched Cube of Shell 2 and 3 and even into Shell 4 is the most utilized as that structure gives clear physical properties – like bonding at a particular angle.

For the larger molecules that properties have smaller differences, and less specific bonding. They have a different structure – a metal interaction which is called alloy.

When a layer is partially full enough, then some positions get filled by the outer electrons from another atom – when it can without disrupting the basic structure of that other atom. That atom sits further away because it has both electrons and protons, but it fills that energy opening, and blocks other atoms. Together, that creates multi-atom molecules.

In the attached tables, we will use the following Legend.

Nucleus, Its Magnetic Field, Electrons, and Bonding Positions

- ◊ Nucleus Position
- ● Electron Position - Generally
- ◈ Electron Position - When in back section of rotation
- ○ Bonding Position

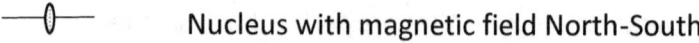   Nucleus with magnetic field North-South

It then has various representations based upon the view direction:

Polar View Based from One Magnetic Pole

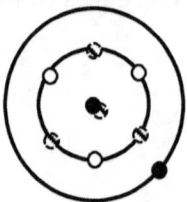

The challenge is that you usually have two in the center (and the 2nd is hidden). That is why you need to review all the directions to get a complete understanding of the atom structure. Understanding the hidden feature usually means looking at one of the other view.

Equatorial View Based Upon the Magnetic Poles

Most of which then fit the oval 1+cos-squared magnetic field. This is a view which expresses around the line (magnetic axis). Therefore, it is more like a 'bagel'.

It will have the magnetic axis line and potentially angle lines. It will have particles in various shells and the nucleus in the center, of course. However, it is a 2D representation of a 3D structure.

Equator View

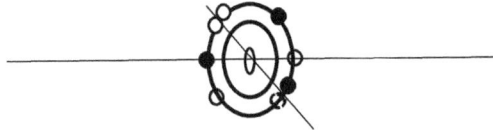

Structure View

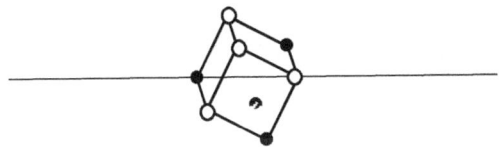

Trying to show 3D in 2D creates challenges. Specifically, we can have some locations front and back. In those cases, this view represents these as two marks touching to an angle line, but that is not really touching, electrons cannot touch each other; instead, it is that the electron sit front and back at the same angle. From the polar view, those are separated, even though they look like they touch as the same angle for the Equatorial View.

Polar View of the Two that Seem to 'Touch'

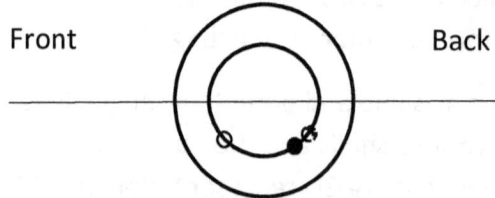

Front   Back

Yet, really those are front and back from the Structure View

Structure View

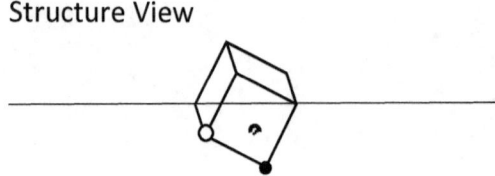

Yet, from the polar view the bonding location would be in the back.

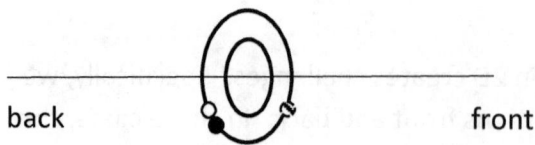

back   front

Shells are shown as ovals as the strength of the magnetic repulsion field towards the equator is strongest, so electrons in the same shell actually sit further away as it moves towards the equator.

This corresponds to the large picture of a magnetic field which is similar to an oval within a certain frame. Again, sort of 'bagel' versus a perfect sphere.

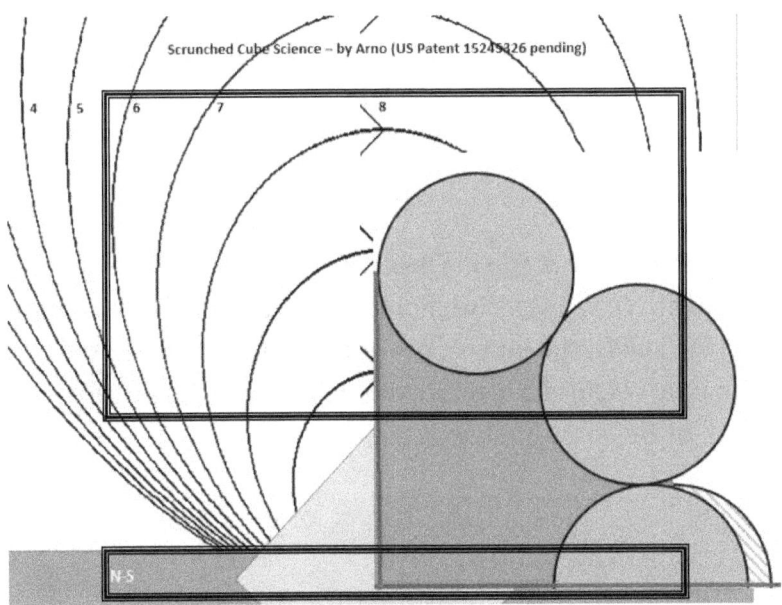

Better picture might have the bagel, but electrons do not sit the part that is different. Therefore, the oval handles the actual items just as well as needed. The reason I just use an oval is that we have an inner shell that already is in that position.

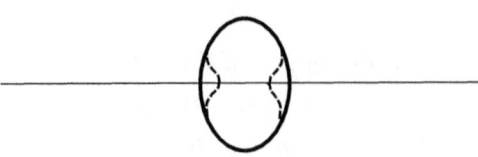

The reason I just use an oval is that we have an inner shell that already is in that position. Therefore, just for a pretty picture, we ignore the full 'bagel' and use ovals for the Equatorial View.

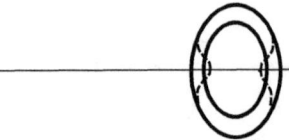

The electrons sit at the magnetic axis and on the outside of the bagel. That inward section of the magnetic field driving electron shells, the indent, never gets an electron, so the use of an oval versus a bagel is necessary so that the picture has room for the inner shells.

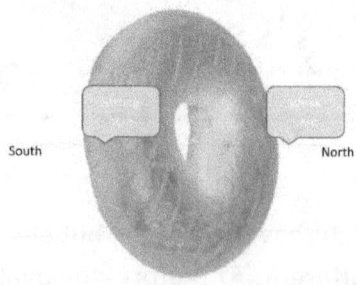

For one structure, 06-C Carbon, the charge repulsion exceeds the magnetic field at the Shell 2 distance, so the structure is a pure sphere (and therefore, a tetrahedron for electrons at Shell-2, and Cube in total, and then the remaining tetrahedron when viewing the four bonding positions). That shells will get shown as a circle representing a sphere when rotated on the magnetic axis.

06-C Carbon – 2nd Shell is spherical

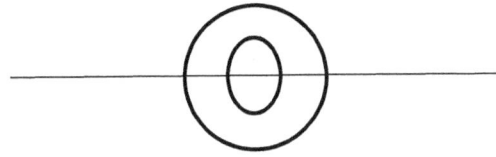

Angles Shown for only the Outer Shell most released electrons or bonds

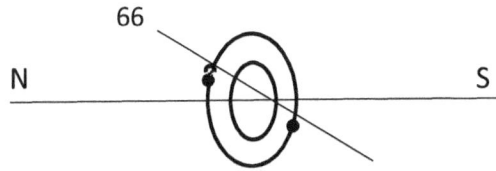

A line with a number shows the angle at the nucleus between the magnetic pole and the electrons. For convenience, a front and back just touch at the line, but physically are front and back at that same angle relative to the magnetic orientation.

- o Electrons in that quadrant show on the line

- o Electrons that would be front and back touch each at the line. The black being front and the hash electron being in the back at the same angle.

Structure View Based Upon the Magnetic Poles

Another view looks at the structure of the outer shell. It has the following possibilities which change based upon the number of shells. As you get bigger, the next inner layer, and spacing creates movement:

Shell 1 – Along Axis – Polar View

For the first shell, the two electrons sit only along the axis, one at north, and the other at south.

Please note that one electron is at the other end, and not visible. I remain undecided if there is view that shows the back electron when it is there like this. The hash ones are in the back.

Shell 2 and Shell 3 – Scrunched Cube

When the outer shell if #2 or #3, the structure is a scrunched cube. However, the electrons get placed at the lowest energy positions of a sphere based upon whether a lower layer exists. It is better to fit at the middle when a lower layer exists.

The lowest energy in 3D is 4 points, a regular tetrahedron. However, that does not have polar symmetry. You cannot cross two positions (electrons) and have the structure balanced along the axis. Therefore, you end up with two overlapping tetrahedrons 2 x 4 = 8 which is a cube. A cube is eight balanced points in regular spherical space.

As the shells build, the next layer wants to fix into the open space of the inner layer. In that sense, Shell 2 is the eight corners of the cube, but Shell 3 is the faces of the cube plus the low energy magnetic poles.

Shell 2 – Corners

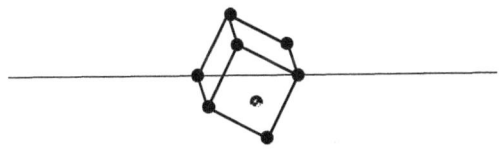

Shell 3 – Ends and Faces

Showing electrons sitting just the face (visible only)

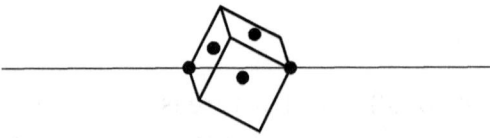

Adding the back electrons looks like:

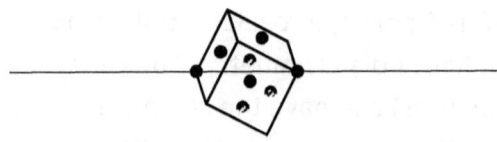

Of course, because of magnetics in near full shells, there is scrunching.

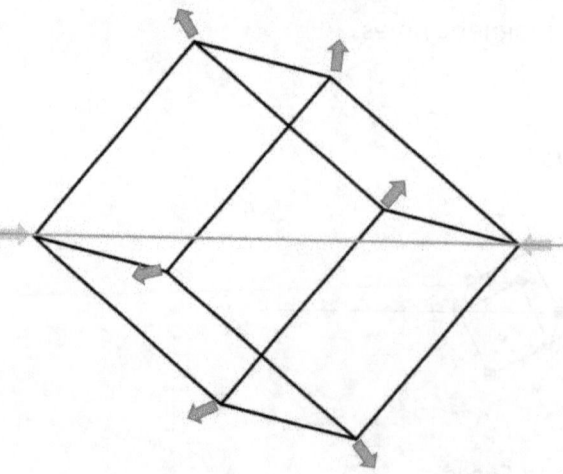

Shell 4 and Shell 5 – 1,3,5 Sphere

Really, this is a representation of a spherical shell with energy strong at the equator. However, electrons are really all over spacing as best they can. The picture is 2D of 3D so the hidden ones take extra time to understand.

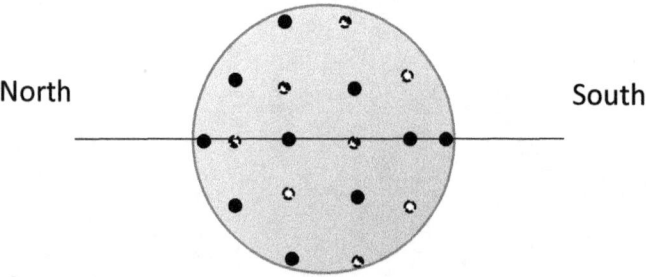

North   South

They build as underlying layers requires getting bigger, and there is room between the electrons themselves.

1-3-5-5-3-1

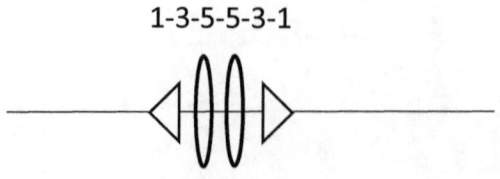

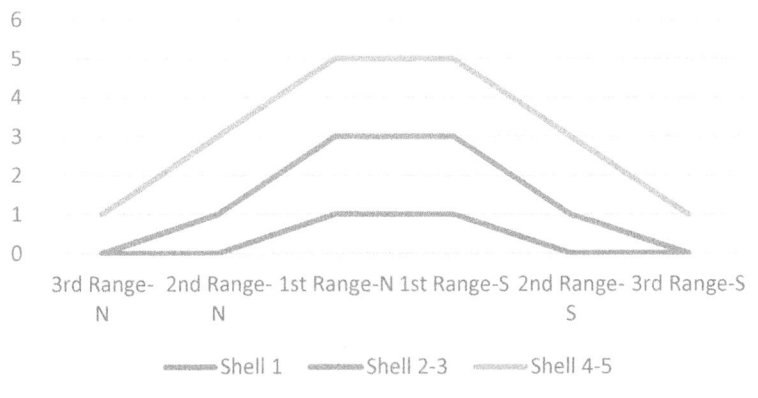

For the outer shell of Shell 4 and Shell 5, the structure is spherical steps of 1,3,5 at each end – so 9 x 2 hemispheres or 18 total electrons in those outer shells.

When only a few (<9) are in a new shell, then only the endcaps are filled. In those case, we will show the center as the prior layer. For Layer 4 that is a scrunched cube. To emphasize that the positions are spread at endcap distances, a dashed interior structure displays.

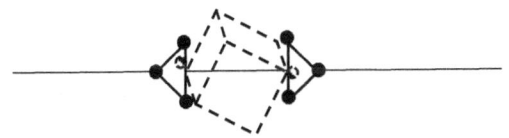

The filling process is from the poles towards the equator. That follows the strength of the magnetic field. That order is from close the magnetic axis 1, to further angles away from the access.

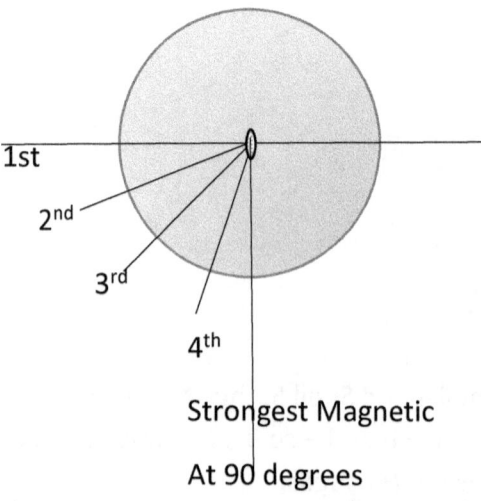

1st
2nd
3rd
4th

Strongest Magnetic
At 90 degrees

Remember the magnetic repulsion strength increases by the angle from the magnetic poles:

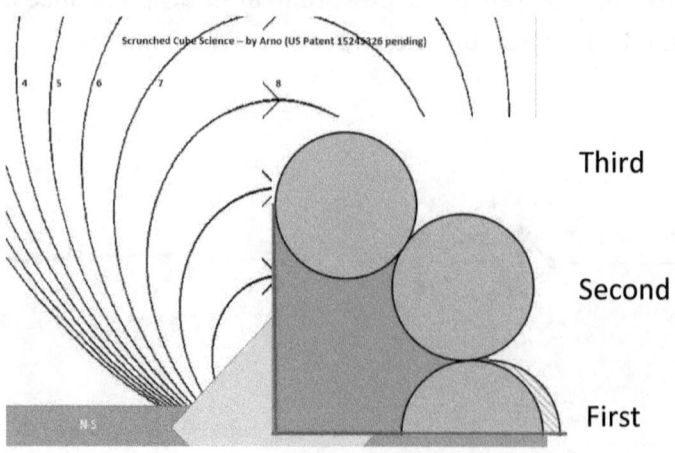

Third
Second
First

Shell 6 and Shell 7 – 1,3,5,7 Sphere

However, there is a different pattern to the filling in Shell 6 and Shell 7. The distances are so big that these two layers fill slightly differently. The magnetic field at 2/distance-cubed has such little influence that by Shell 7, the filling is almost strictly spherical.

For Shell 6, there is an intermediate section where the Range 2 fills to seven, but then backs down to five (5) as the Range 1 gets used.

Shell 6

Range (in order)	Intermediate	Full Shell
Range 4 - Polar	1	1
Range 3	3	3
Range 2	7	5
Range 1 – Near Equator	0	7

However, for Range 7, that does not happen. In fact, because of the Shell-6 layer underneath, this layer fills much more diversely with Intermediate layers

Shell 7

Range (in order)	Intermediate	Full Shell
Range 4 - Polar	1	1
Range 3	1	3
Range 2	1	5
Range 1 – Near Equator	1	7

This is very different logic than the Pauli-Aufbau filling logic. It is very specific to the 3D Geometry which takes into account the next-inner layer repulsions. For even layers, the lower structure is different in size and shape, so it is 'lumpy'. For odd layers, the lower layer is the same shape and structure so those odd layers generally just offset by half the angle to 'fill in the open space'.

For the outer shell of Shell 4 and Shell 5, the structure is spherical steps of 1,3,5,7 at each end – so 9 x 2 hemispheres or 18 total electrons in those outer shells.

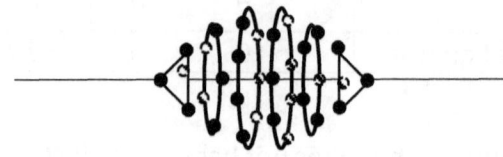

Now, the ends are scrunched so this actually is a scrunched sphere when properly spaced.

From the polar direction, it looks like:

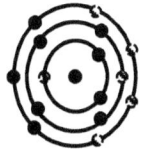

Again, this is a pattern based upon more electrons fit as the perimeter increases. It goes from pole to pole.

1,3,5,7,7,5,3,1

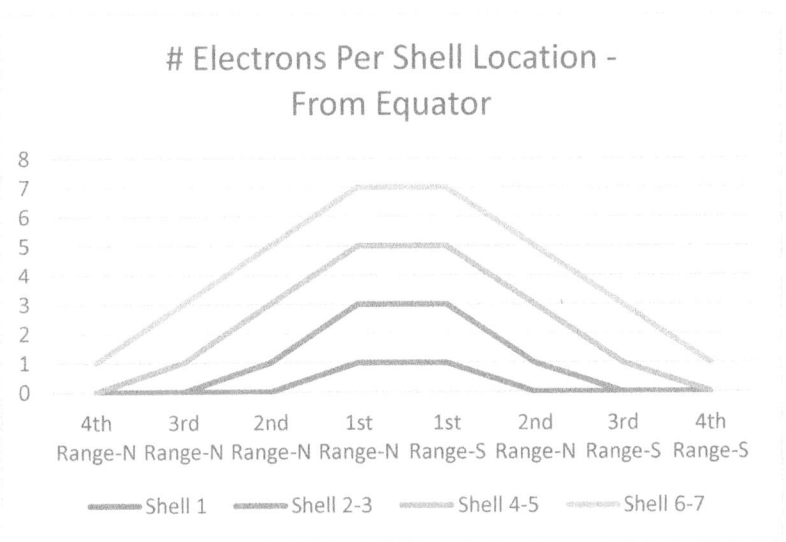

Also note, that to fit at the lowest energy, the two hemispheres are offset by 360/N/2.

Therefore, at Shell 4 and Shell 5, the north side is offset 360/5/2 degrees of latitude = 36 degrees. That makes the centers fits best, then the other longitude-rows offset based upon that interface.

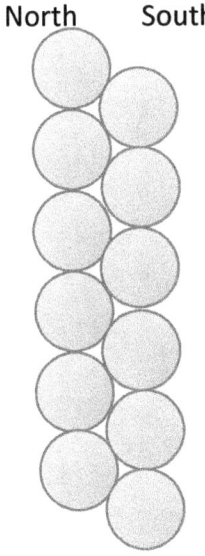

Wrapping back to the top

Organizing from 7 to 5 to 3 to 1 is based upon a sphere, so it looks like:

1/3/5

Black – 5m1

Orange – 5t3

Red – 4t3

Blue 5u5

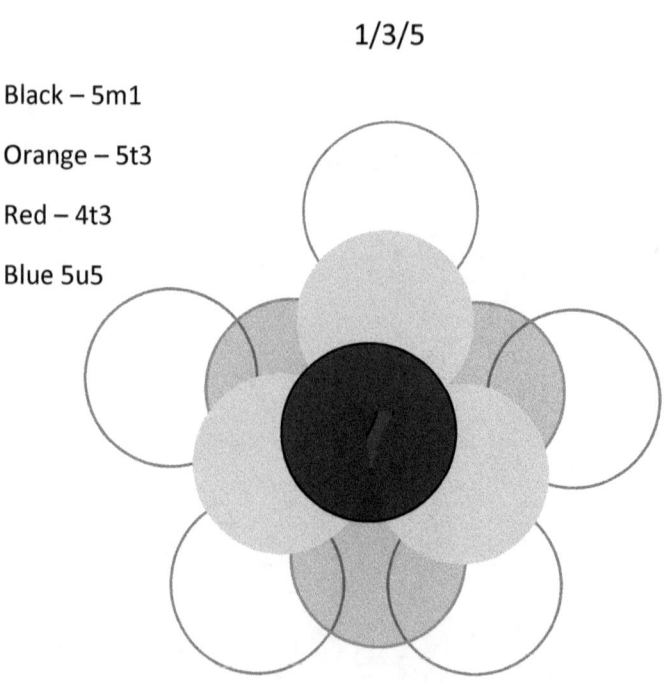

Electrons Shell Configuration and Bonding Locations of Chart of Elements

001-H Hydrogen

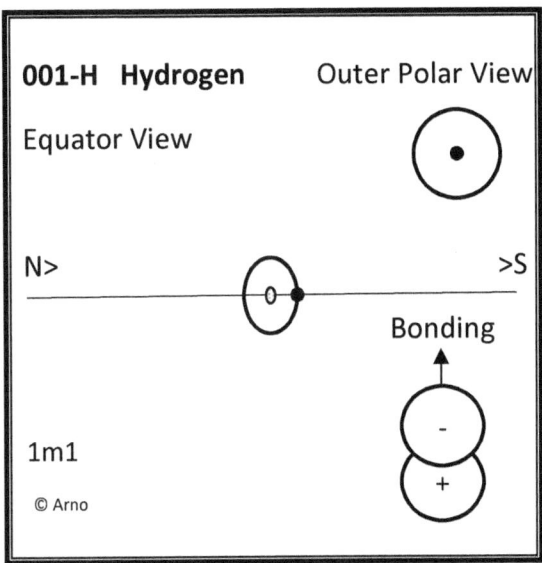

Shell/Subshell/Count	# Electrons	Structure
1m1	1	Magnetic Polar

001-H Hydrogen – Side View

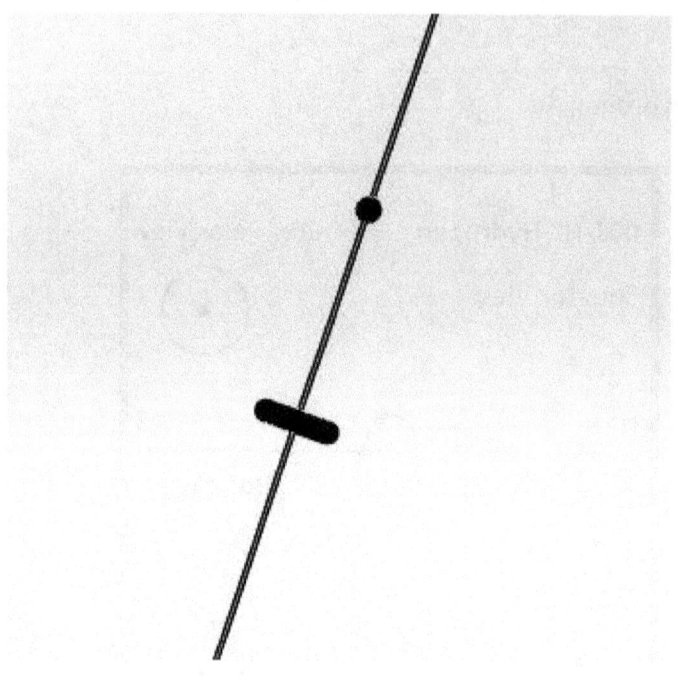

Legend for Mechanical Drawings

Sphere – Electron

Ring - Nucleus

Line – Magnetic (not an actual connector, so particles float, oscillate, and rotate/orbit within stable structure (not an actual connector, so particles float, oscillate, and rotate/orbit within stable structure)

Small Line – Electron-electron repulsion

001-H Hydrogen – Isometric View – Bonding Electron Facing Towards Reader

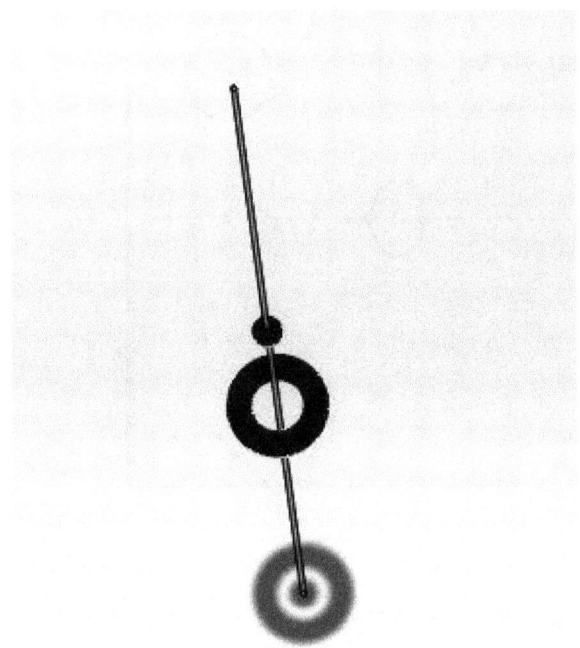

002-He Helium

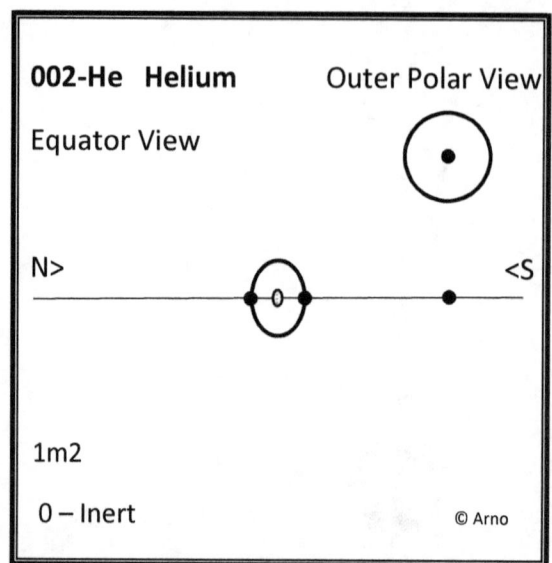

Shell/Subshell/Count	# Electrons	Structure
1m2	2	Magnetic Polar

002-He Helium – Isometric View

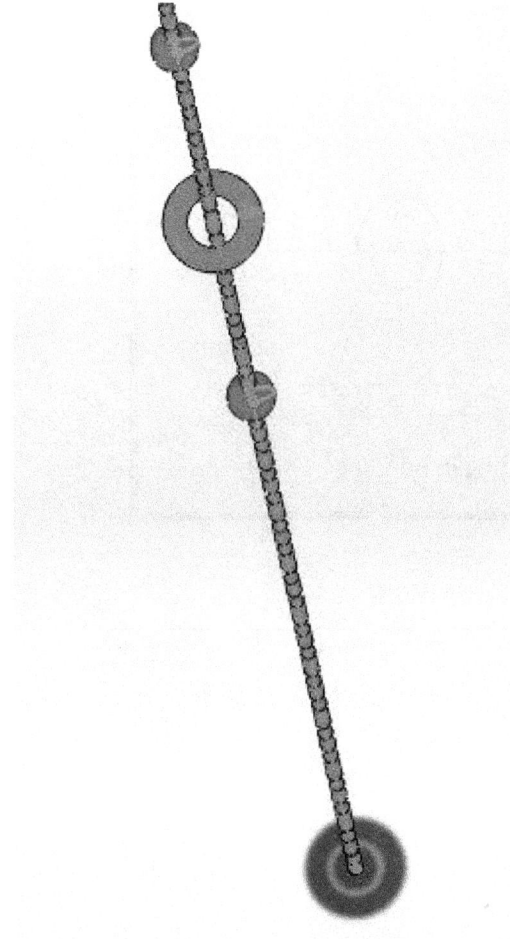

003-Li Lithium

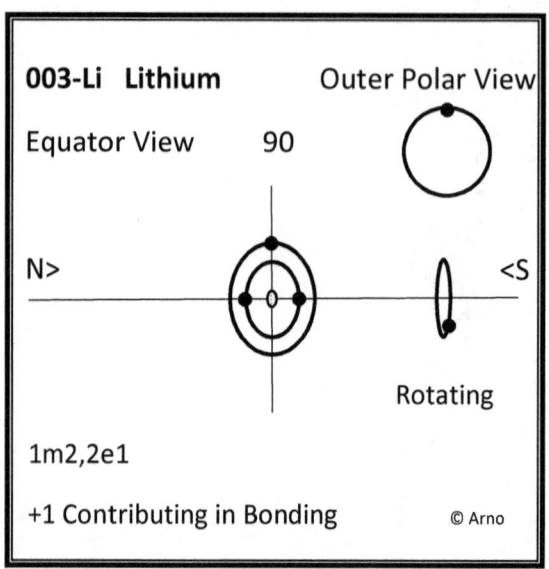

Shell/Subshell/Count	# Electrons	Structure
1m2	2	Magnetic Polar
1e1	1	Equatorial

003-Li Lithium – Equatorial View

This shows that the one electron in 2-Shell does not side at 2m1 (on the magnetic pole), but instead at the equator relative to that magnetic axis.

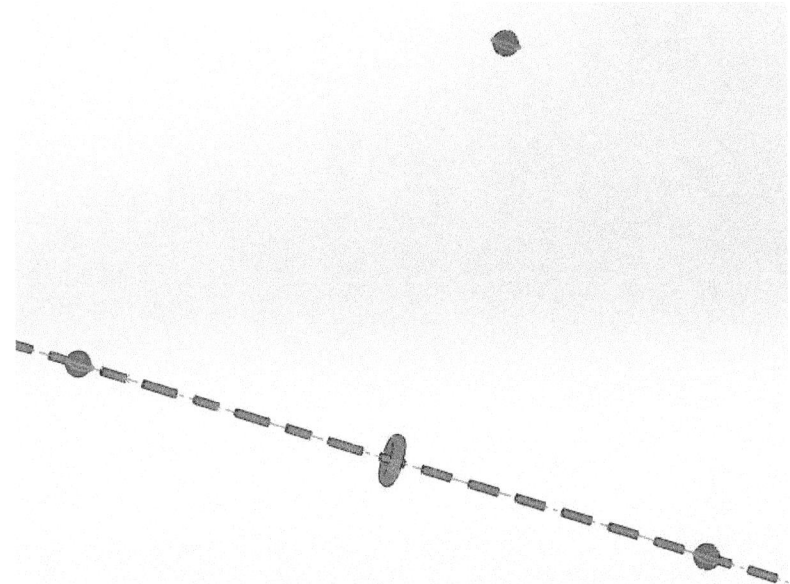

003-Li Lithium – Polar View

Further, it shows that from the end. The electron is wide (at the equator) of the e-2m1/nucleus/e-2m2 structure

004-Be Beryllium

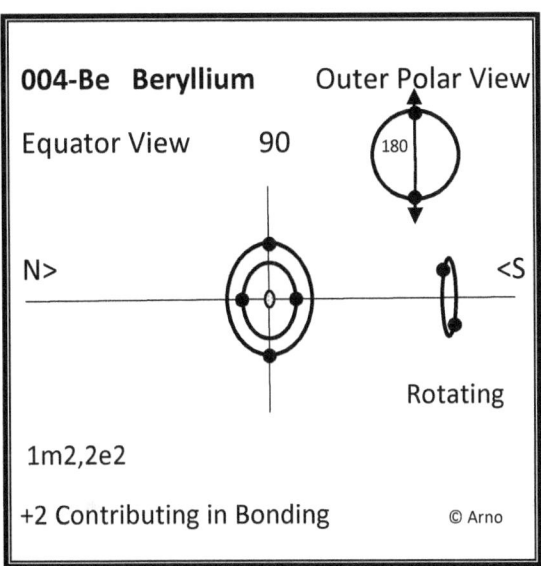

Shell/Subshell/Count	# Electrons	Structure
1m2	2	Magnetic Polar
1e2	2	Equatorial

005-B Boron

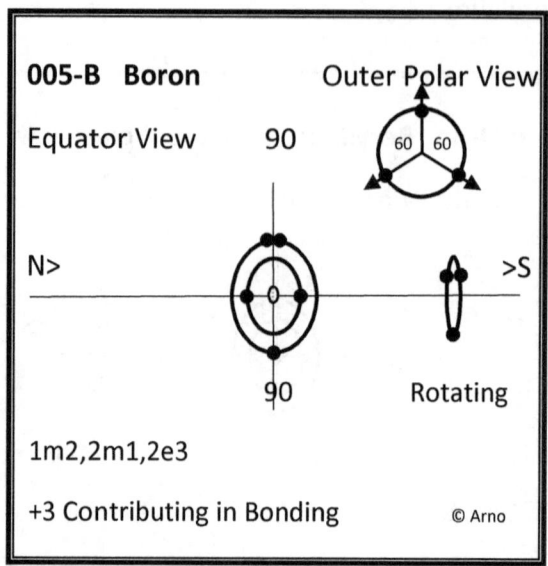

Shell/Subshell/Count	# Electrons	Structure
1m2	2	Magnetic Polar
1e3	3	Equatorial

005-B Boron – Side/Equator View

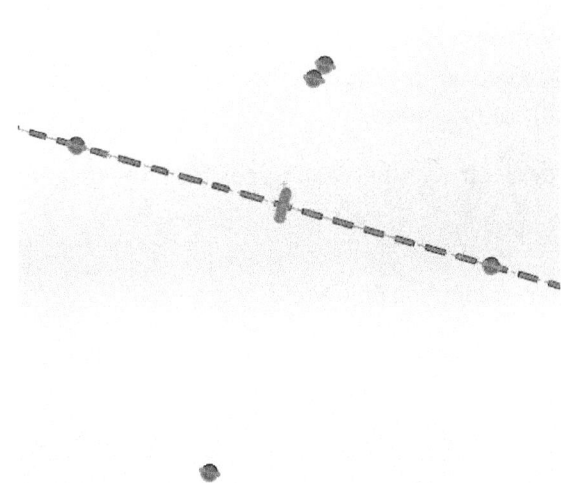

Isometric View

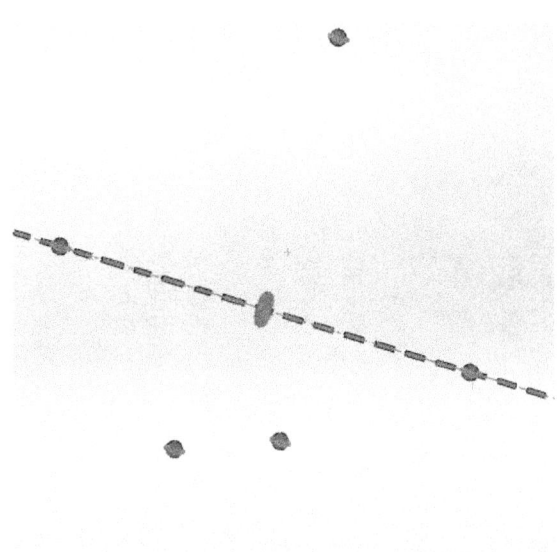

Boron follows one of the equatorial exceptions. Because there are not lower layers, the 2m shell is not filled first. Instead, from up to 2e3, the electron seeks a direct equatorial placement. Above 4 in the 2-Shell, electrons location return to 2m2 with the 2c (corners of the cube) configuration.

The equator electrons sit at 90 degrees from the magnetic axis.

005-B Boron – Magnetic Pole View

From the polar view, you can see the 3-electron sit at 120 degrees equally around the magnetic axis.

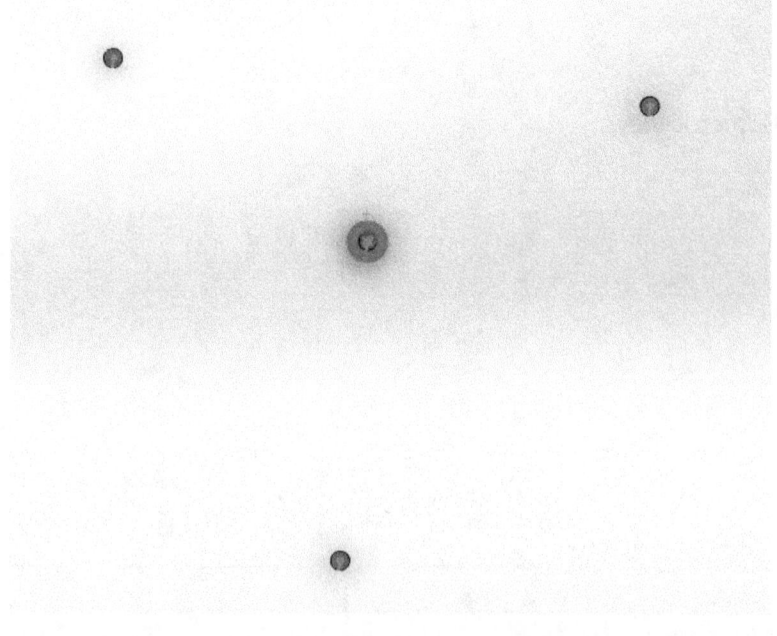

006-C Carbon

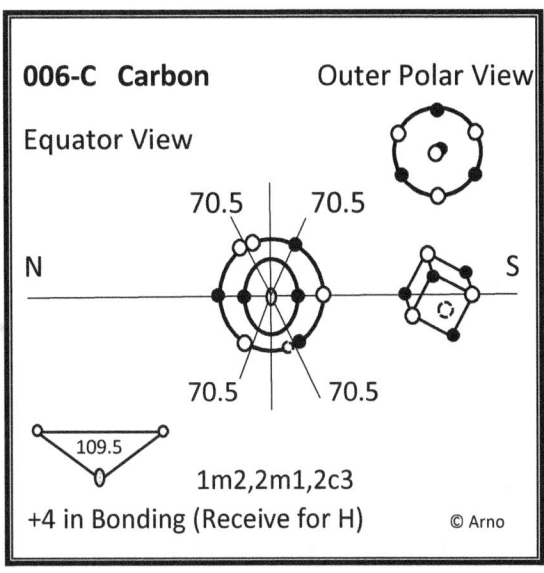

Shell/Subshell/Count	# Electrons	Structure
1m2	2	Magnetic Polar
2m1	1	Magnetic Polar
1c3	3	Tetrahedron-1/2 Cube

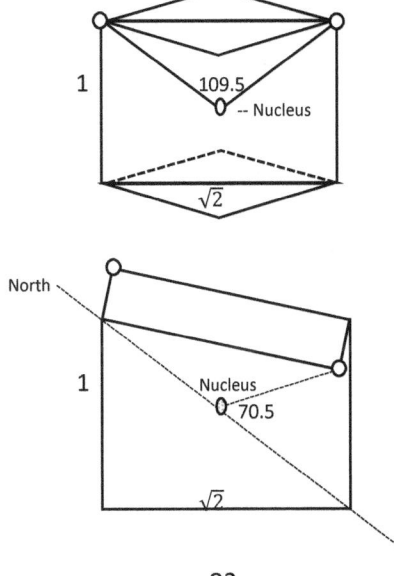

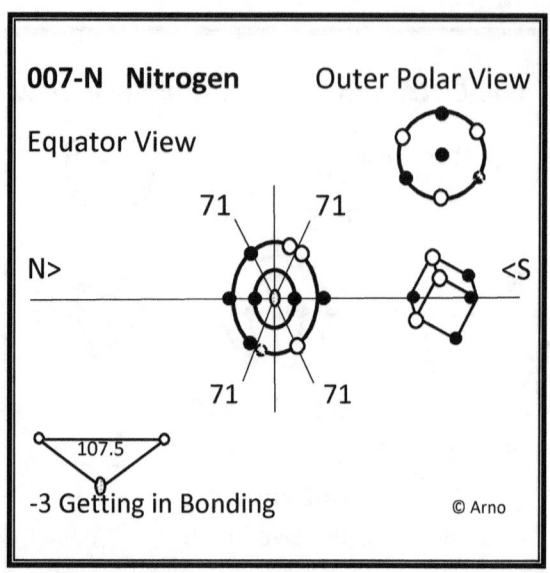

Shell/Subshell/Count	# Electrons	Structure
1m2	2	Magnetic Polar
2m2	2	Magnetic Polar
1c3	3	Scrunched Cube

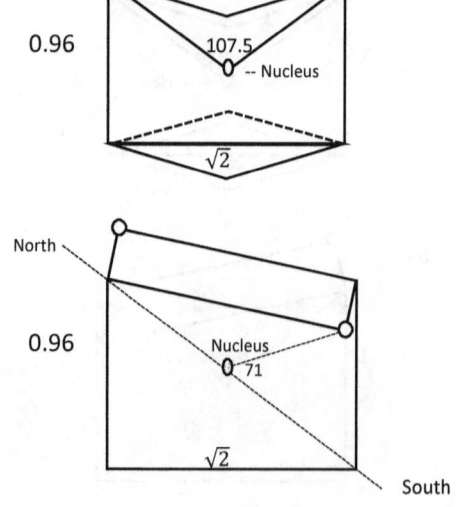

008-O Oxygen

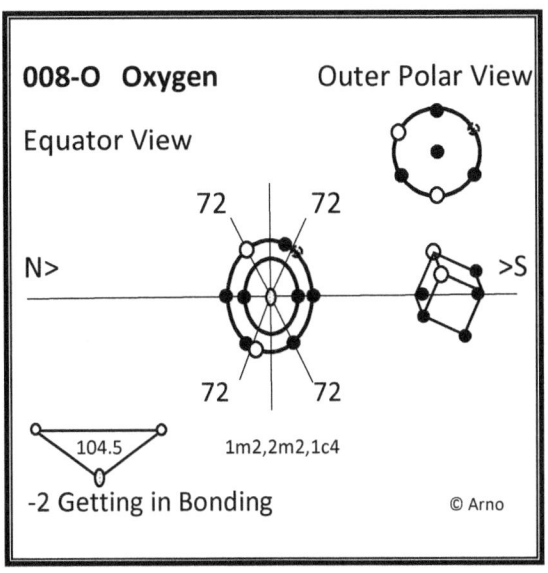

Shell/Subshell/Count	# Electrons	Structure
1m2	2	Magnetic Polar
2m2	2	Magnetic Polar
1c4	4	Scrunched Cube

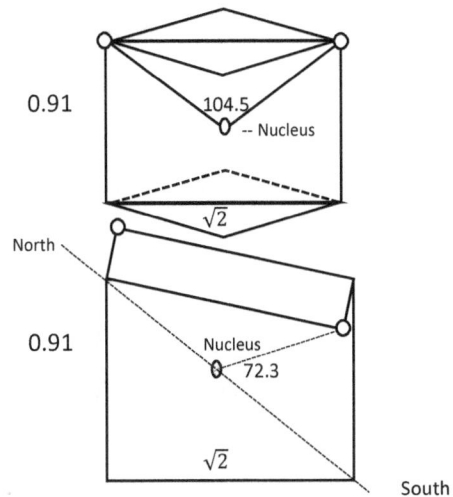

009-F Florine

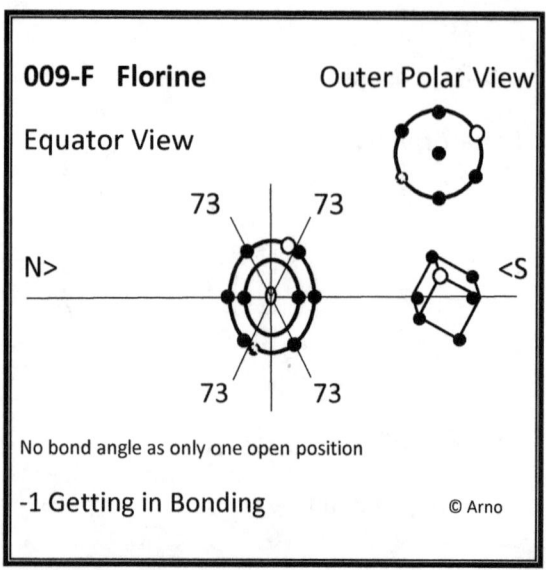

Shell/Subshell/Count	# Electrons	Structure
1m2	2	Magnetic Polar
2m2	2	Magnetic Polar
1c5	5	Scrunched Cube

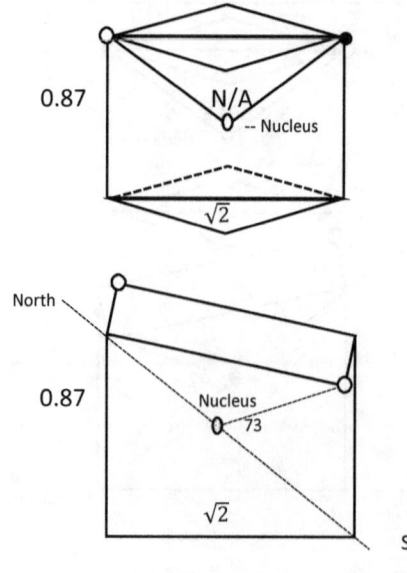

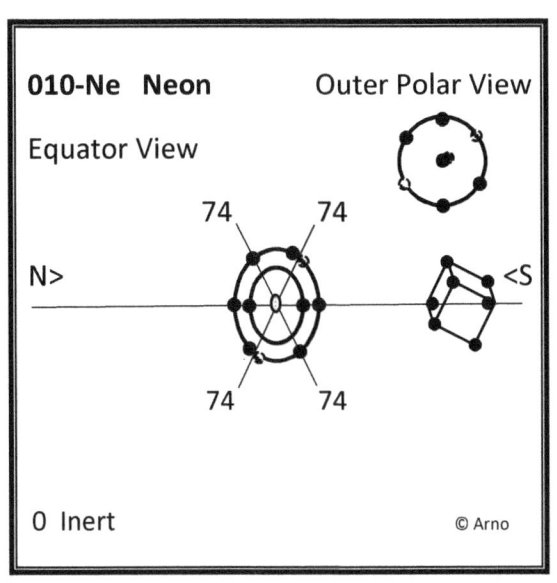

Shell/Subshell/Count	# Electrons	Structure
1m2	2	Magnetic Polar
2m2	2	Magnetic Polar
1c6	6	Scrunched Cube

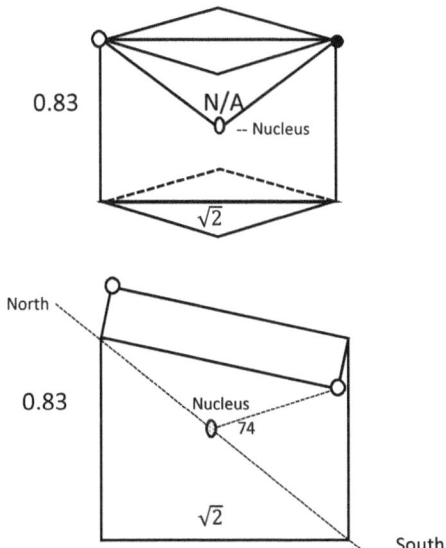

011-Na Sodium

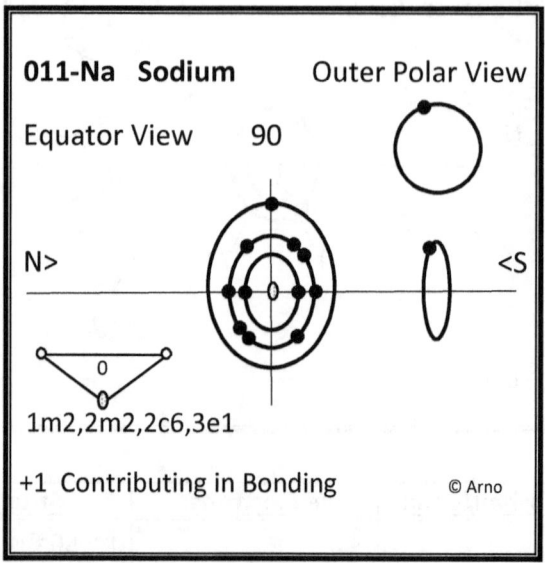

Shell/Subshell/Count	# Electrons	Structure
1m2	2	Magnetic Polar
2m2	2	Magnetic Polar
1c6	6	Scrunched Cube
2e1	1	Equatorial

+

012-Mg Magnesium

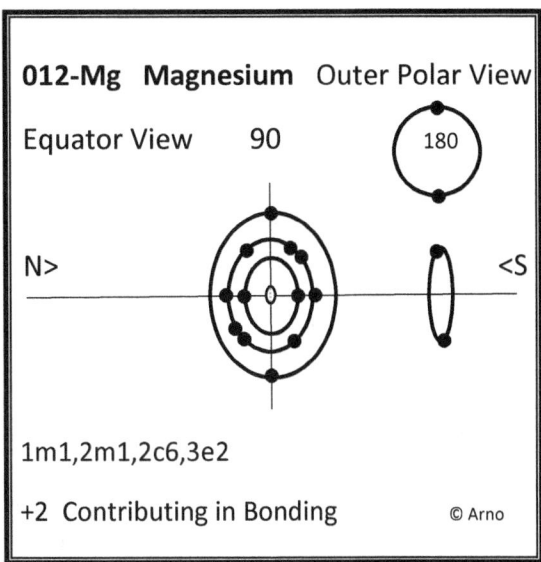

Shell/Subshell/Count	# Electrons	Structure
1m2	2	Magnetic Polar
2m2	2	Magnetic Polar
1c6	6	Scrunched Cube
2e2	2	Equatorial

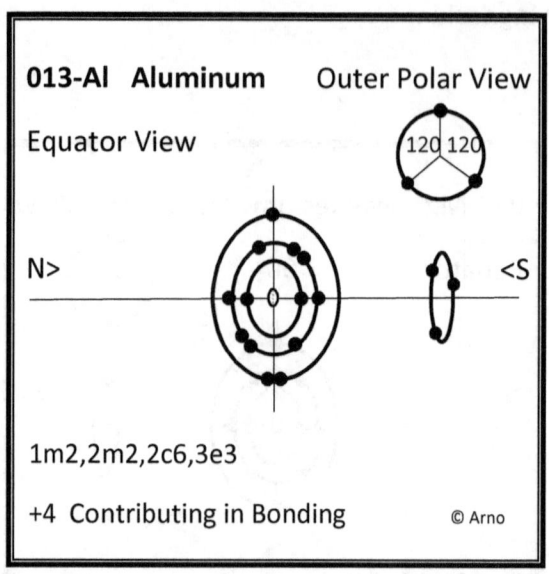

Shell/Subshell/Count	# Electrons	Structure
1m2	2	Magnetic Polar
2m2	2	Magnetic Polar
1c6	6	Scrunched Cube
2e3	3	Equatorial

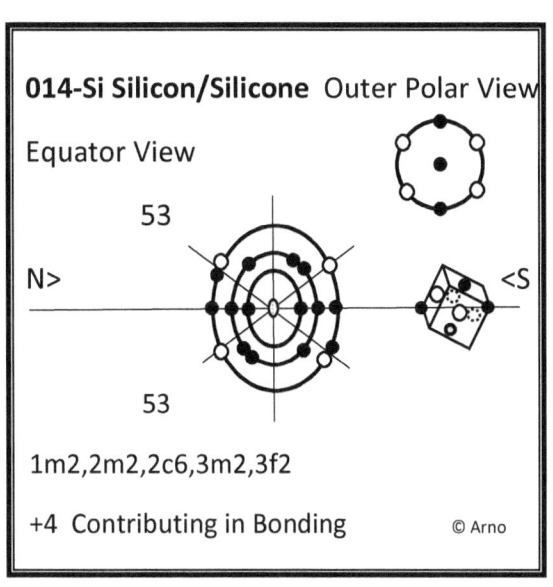

Shell/Subshell/Count	# Electrons	Structure
1m2	2	Magnetic Polar
2m2	2	Magnetic Polar
1c6	6	Scrunched Cube
2m2	2	Magnetic Polar
2f2	2	Faces of the Cube

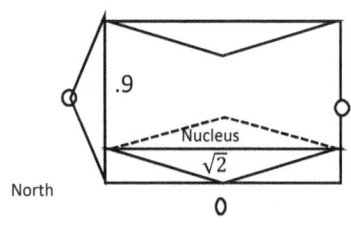

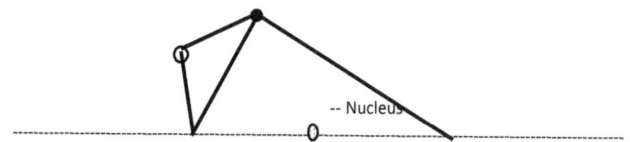

Silicone, because it sits in the Shell-3 faces structure, remains one of few elements that have electron filling and open bonding locations located at about 90 degrees (there is a slight tweak to this because of the repulsion from the magnetic pole e-3m2 electrons). Further, Silicone is the only elements with four open locations all at 90 degrees in a ring.

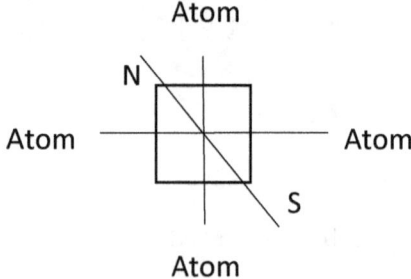

It is no wonder that silicon use in computers 0 and 1 on off works so much better than other elements with non-90 degree bonds. That makes the two filled electrons (e-3f2) well isolated a) at 180 degrees, and b) the four bonds

This bonding at Shell-3 face structure also creates the reasoning for silicone crystal structure to be face-centered diamond-cubic.

While it falls in the column of 006-C Carbon above, in the Shell-2 corners of the cube, Silicone, because of the faces from the lower layers, has few properties in common with elements both above and below it. The angles are 90 versus 109.5; a lump of coal is not at all like the jelly of a blob of silicone.

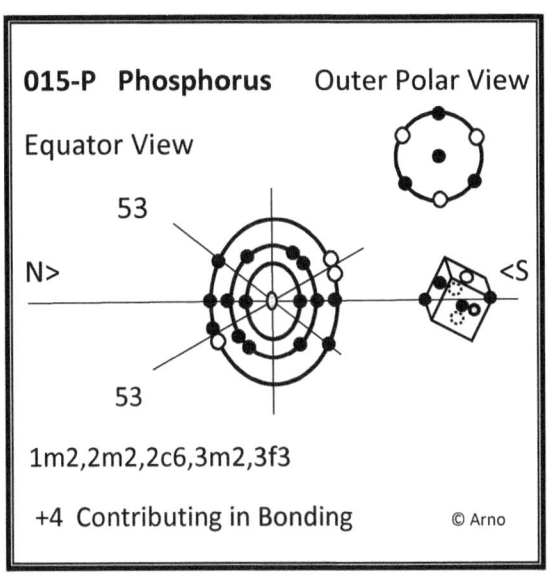

Shell/Subshell/Count	# Electrons	Structure
1m2	2	Magnetic Polar
2m2	2	Magnetic Polar
1c6	6	Scrunched Cube
2m2	2	Magnetic Polar
2f3	3	Faces of the Cube

Phosphorus is unique in that the three bonding locations are a) aligned at 90 degrees, and 180, and b) sit in steps around the faces. The electrons filled have that same 90, 180 construction.

Most bonding positions are >90 degrees, meaning that other atoms are further apart. Therefore, 015-P tends to have one of bonds more likely to break – and create light or to make sequential jumps from bonding in one 90-degree slot into another.

016-S Sulfur – Common -2 Configuration

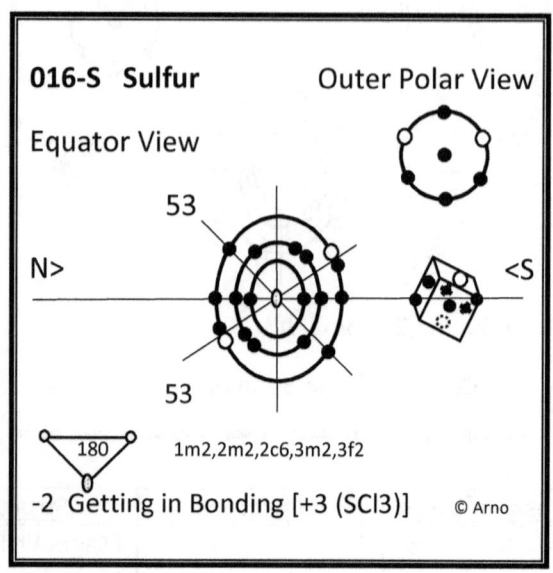

Shell/Subshell/Count	# Electrons	Structure
1m2	2	Magnetic Polar
2m2	2	Magnetic Polar
1c6	6	Scrunched Cube
2m2	2	Magnetic Polar
2f4	4	Faces of Cube

016-S Sulfur – +6 (+3 x Double) Configuration (SO3)

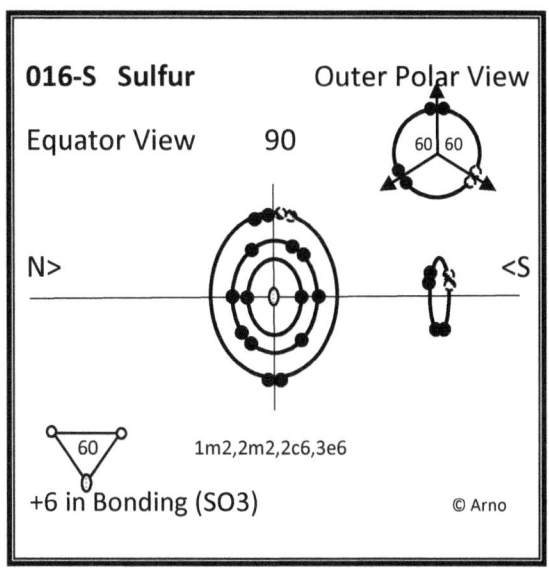

Shell/Subshell/Count	# Electrons	Structure
1m2	2	Magnetic Polar
2m2	2	Magnetic Polar
1c6	6	Scrunched Cube
2m2	0	Magnetic Polar
2e6	6	Equatorial

017-Cl Chlorine

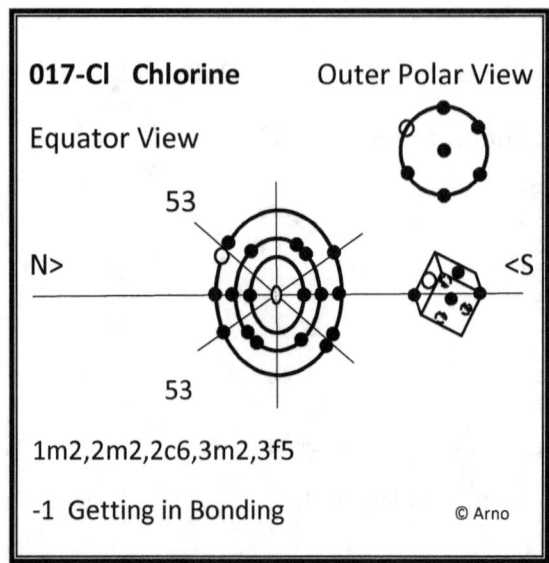

Shell/Subshell/Count	# Electrons	Structure
1m2	2	Magnetic Polar
2m2	2	Magnetic Polar
1c6	6	Scrunched Cube
2m2	2	Magnetic Polar
2f5	5	Faces of Cube

018-Ar Argon

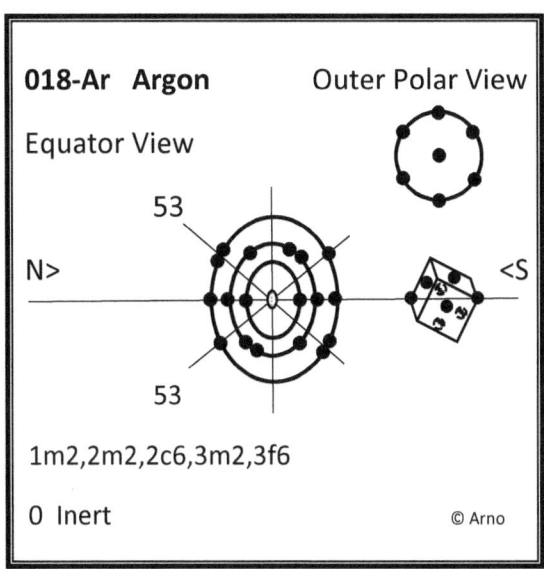

Shell/Subshell/Count	# Electrons	Structure
1m2	2	Magnetic Polar
2m2	2	Magnetic Polar
2c6	6	Scrunched Cube
3m2	2	Magnetic Polar
3f6	6	Faces of Cube

019-K Potassium

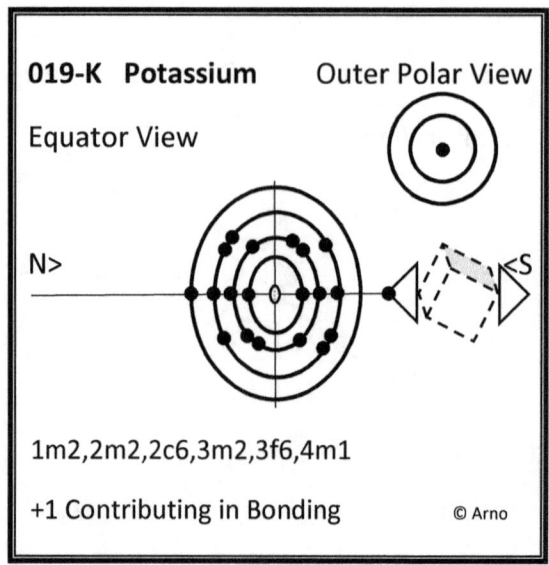

Shell/Subshell/Count	# Electrons	Structure
1m2	2	Magnetic Polar
2m2	2	Magnetic Polar
2c6	6	Scrunched Cube
3m2	2	Magnetic Polar
3f6	6	Faces of Cube
4m1	1	Magnetic Polar

020-Ca Calcium

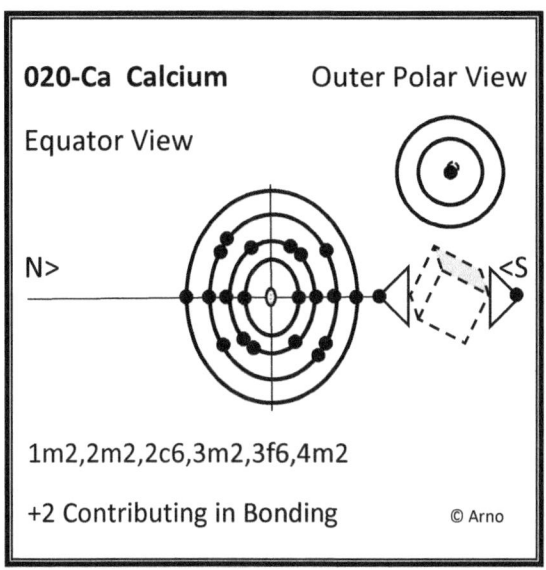

Shell/Subshell/Count	# Electrons	Structure
1m2	2	Magnetic Polar
2m2	2	Magnetic Polar
2c6	6	Scrunched Cube
3m2	2	Magnetic Polar
3f6	6	Faces of Cube
4m2	2	Magnetic Polar

021-Sc Scandium (##OR IS THIS 90 DEGREES)

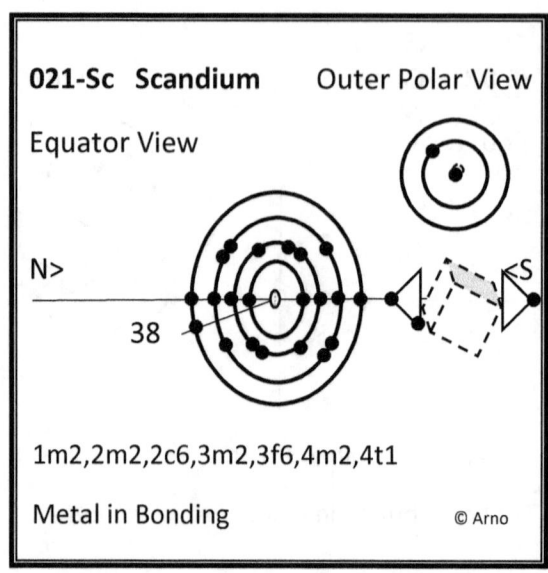

Shell/Subshell/Count	# Electrons	Structure
1m2	2	Magnetic Polar
2m2	2	Magnetic Polar
2c6	6	Scrunched Cube
3m2	2	Magnetic Polar
3f6	6	Faces of Cube
4m2	2	Magnetic Polar
4t1	1	Endcap Tetrahedron

022-Ti Titanium

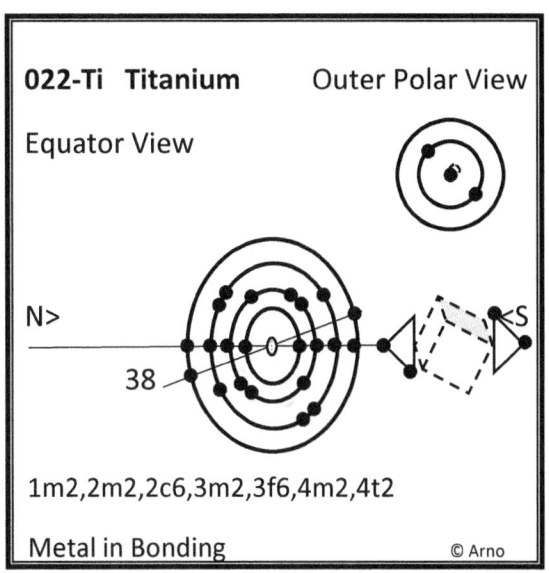

Shell/Subshell/Count	# Electrons	Structure
1m2	2	Magnetic Polar
2m2	2	Magnetic Polar
2c6	6	Scrunched Cube
3m2	2	Magnetic Polar
3f6	6	Faces of Cube
4m2	2	Magnetic Polar
4t2	2	Endcap Tetrahedron

023-V Vanadium

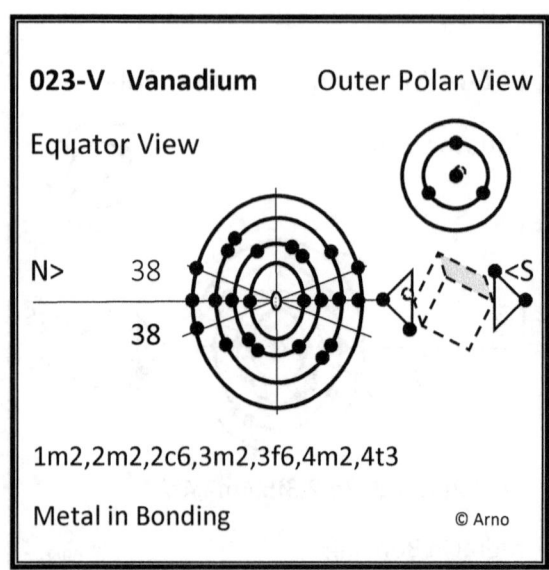

Shell/Subshell/Count	# Electrons	Structure
1m2	2	Magnetic Polar
2m2	2	Magnetic Polar
2c6	6	Scrunched Cube
3m2	2	Magnetic Polar
3f6	6	Faces of Cube
4m2	2	Magnetic Polar
4t3	3	Endcap Tetrahedron

024-Cr Chromium

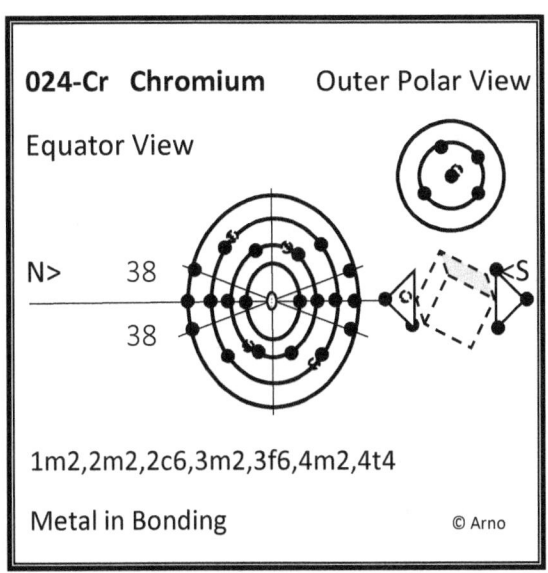

Shell/Subshell/Count	# Electrons	Structure
1m2	2	Magnetic Polar
2m2	2	Magnetic Polar
2c6	6	Scrunched Cube
3m2	2	Magnetic Polar
3f6	6	Faces of Cube
4m2	2	Magnetic Polar
4t4	4	Endcap Tetrahedron

025-Mn Mangenese

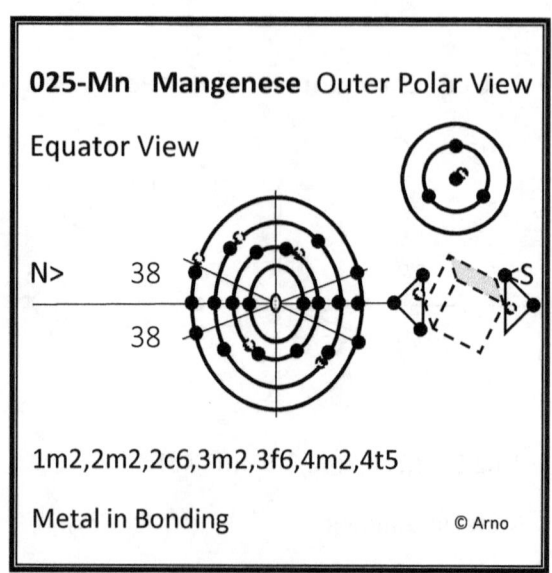

Shell/Subshell/Count	# Electrons	Structure
1m2	2	Magnetic Polar
2m2	2	Magnetic Polar
2c6	6	Scrunched Cube
3m2	2	Magnetic Polar
3f6	6	Faces of Cube
4m2	2	Magnetic Polar
4t5	5	Endcap Tetrahedron

026-Fe Iron

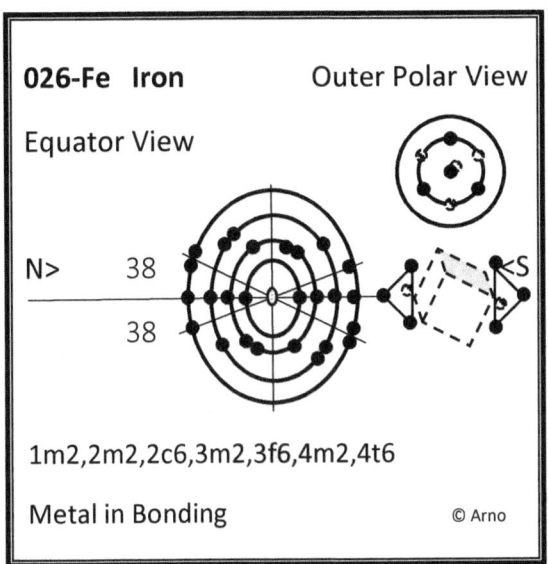

Shell/Subshell/Count	# Electrons	Structure
1m2	2	Magnetic Polar
2m2	2	Magnetic Polar
2c6	6	Scrunched Cube
3m2	2	Magnetic Polar
3f6	6	Faces of Cube
4m2	2	Magnetic Polar
4t6	6	Endcap Tetrahedron

026-Fe Iron – Magnetic Pole View

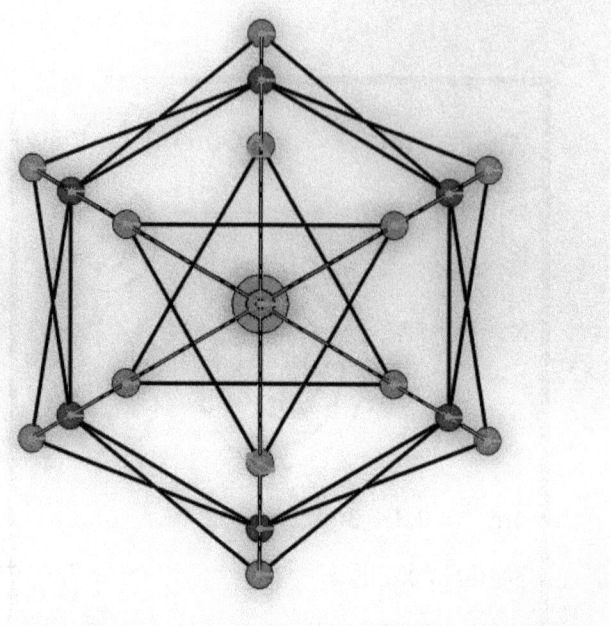

The key properties of 026-Fe Iron occur because all layers of electrons sit in a very balanced structure, also well balanced along the magnetic axis. This provides important properties:

1) Iron is very magnetic, and subject to stabilizing on a magnetic field for extended periods. Look at how balanced the electrons sit around the magnetic centerline.

2) Iron is not electrically conductive. In this 4m2,4t6 endcap structure, no free electron

Further, Iron has exposed electrons from multiple subshells (4m2,4t6,3f6,2c6). All those have easy access to accept and emit radiation. With 22 electrons, the exchange distances between subshells and between electron in the same shell give the large variety of spectrum lines. This electron configuration makes Iron one of the elements with the most diverse, balanced set of wavelengths in spectrum analysis.

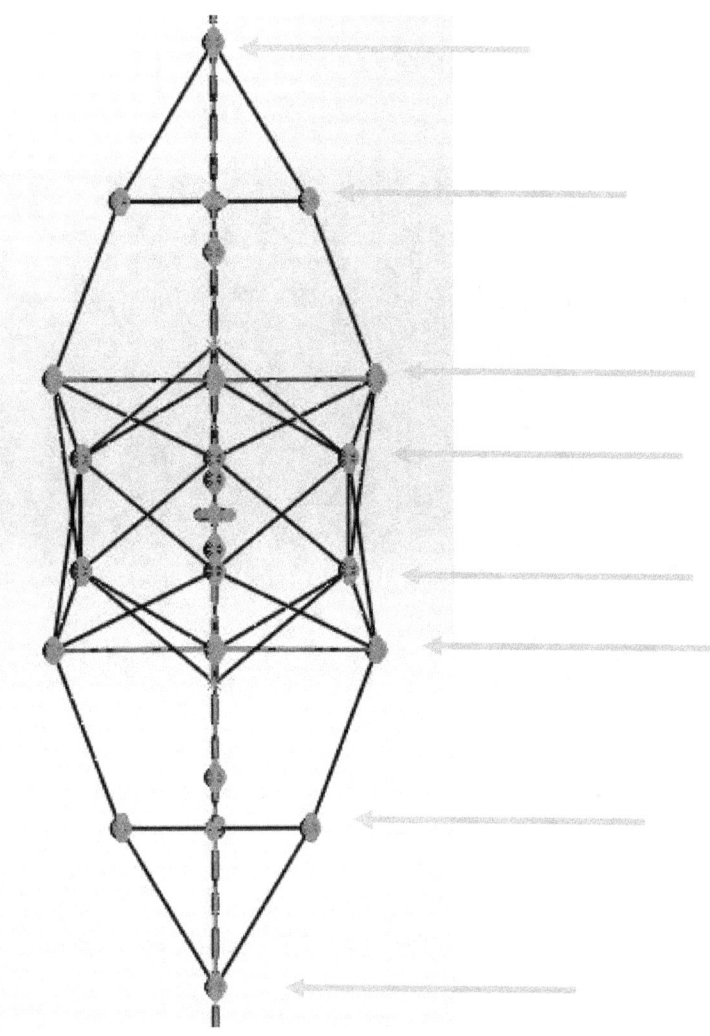

This can get seen in the comparison of Iron spectrum versus its neighbor 027-Co Cobalt in this Period Charge of Spectrum. The number of locations for different exchanges of electrons positions that are exposed to release – at the same horizon area, is significant. You have four sets in each hemisphere, all about equally open to the world.

026-Fe

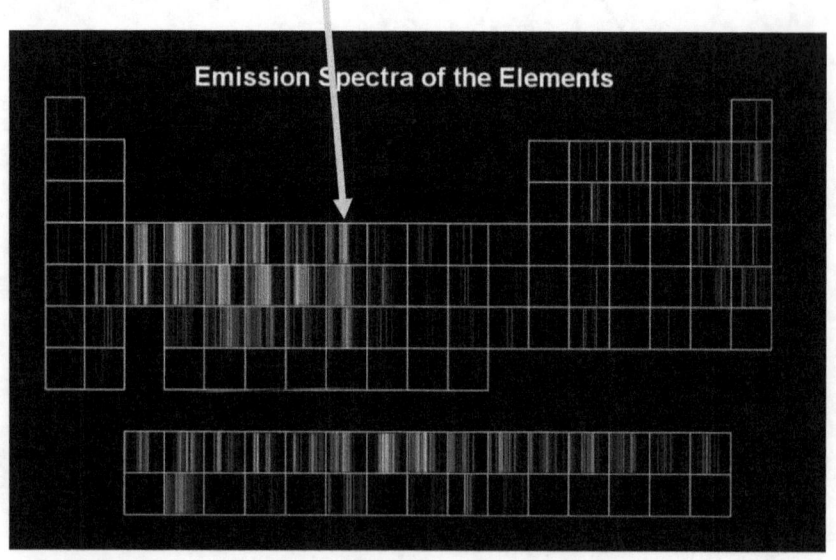

Yet, as you will see in 027-Co next, it has one electron subshell that is substantially a) alone having a huge horizon where to release, and b) substantially more radius to the nucleus (meaning less energy holding it). That makes all position after 8 in large shells drop dramatically in the number of visible, strong spectrum lines.

027-Co Cobalt – Magnetic Configuration

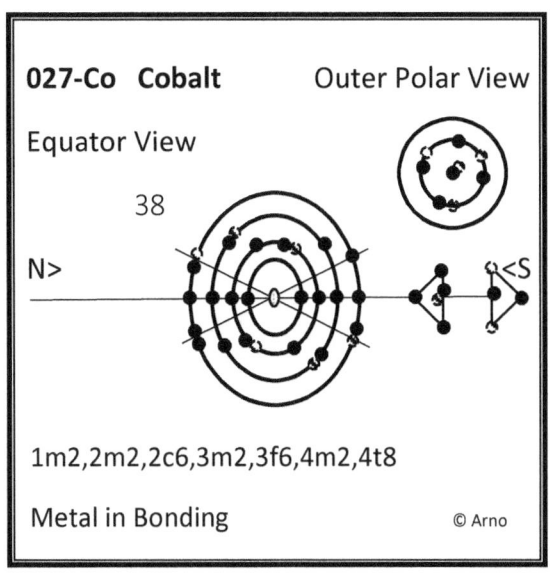

Shell/Subshell/Count	# Electrons	Structure
1m2	2	Magnetic Polar
2m2	2	Magnetic Polar
2c6	6	Scrunched Cube
3m2	2	Magnetic Polar
3f6	6	Faces of Cube
4m2	2	Magnetic Polar
4t6	6	Endcap Tetrahedron
4e1	1	Equatorial

The last electron in Cobalt has two positions depending on the repulsion forces of surrounding atoms and heat. This pages shows where the electron (e-4f7) fills 1/4/0/0/3/1 of 1/3/5/5/3/1, and fits in situations where magnetics applies more. The other electron location (e-4e1) of 1/3/1/3/1 of equatorial 1/4/3/3/1 is the more electrically active configuration, and on the next page.

027-Co Cobalt – Electrical Configuration

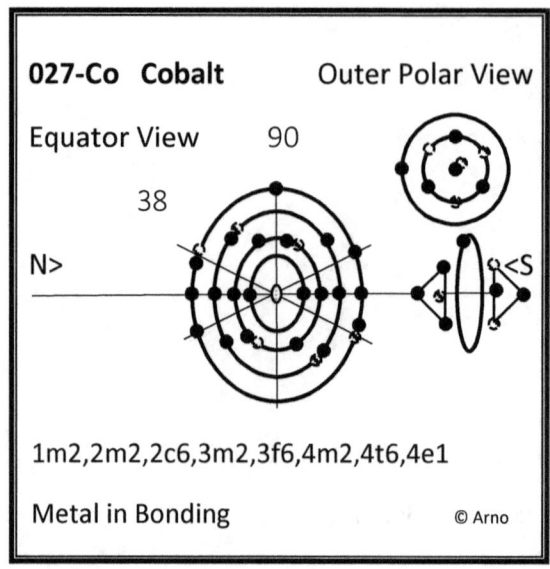

Shell/Subshell/Count	# Electrons	Structure
1m2	2	Magnetic Polar
2m2	2	Magnetic Polar
2c6	6	Scrunched Cube
3m2	2	Magnetic Polar
3f6	6	Faces of Cube
4m2	2	Magnetic Polar
4t6	6	Endcap Tetrahedron
4e1	1	Equatorial

028-Ni Nickel – Magnetic Configuration

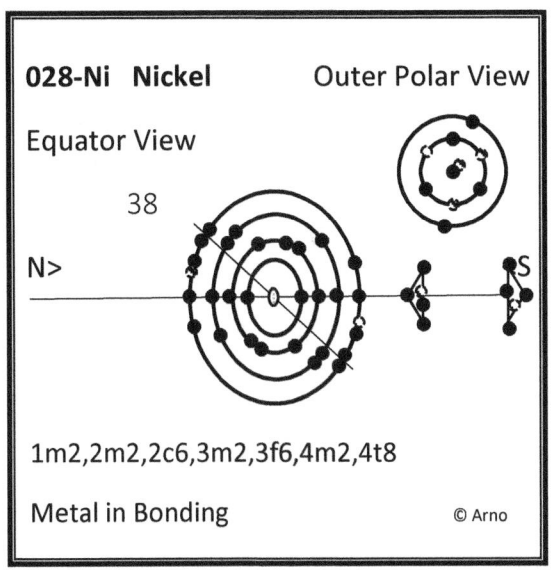

Shell/Subshell/Count	# Electrons	Structure
1m2	2	Magnetic Polar
2m2	2	Magnetic Polar
2c6	6	Scrunched Cube
3m2	2	Magnetic Polar
3f6	6	Faces of Cube
4m2	2	Magnetic Polar
4t6	6	Endcap Tetrahedron
4e2	2	Equatorial

028-Ni Nickel – Electrical Configuration

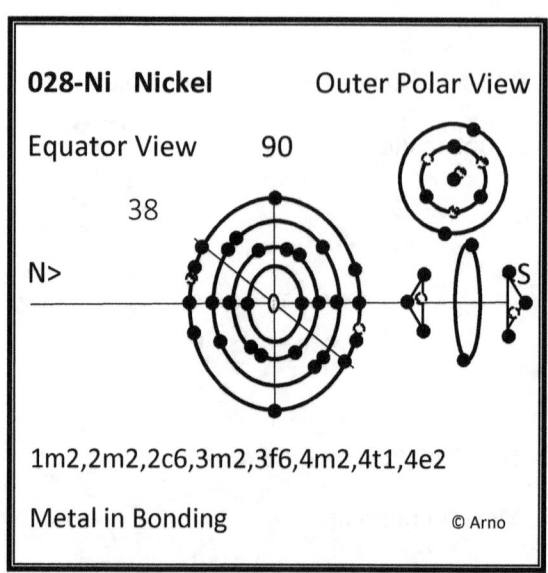

Shell/Subshell/Count	# Electrons	Structure
1m2	2	Magnetic Polar
2m2	2	Magnetic Polar
2c6	6	Scrunched Cube
3m2	2	Magnetic Polar
3f6	6	Faces of Cube
4m2	2	Magnetic Polar
4t6	6	Endcap Tetrahedron
4e2	2	Equatorial

029-Cu Copper

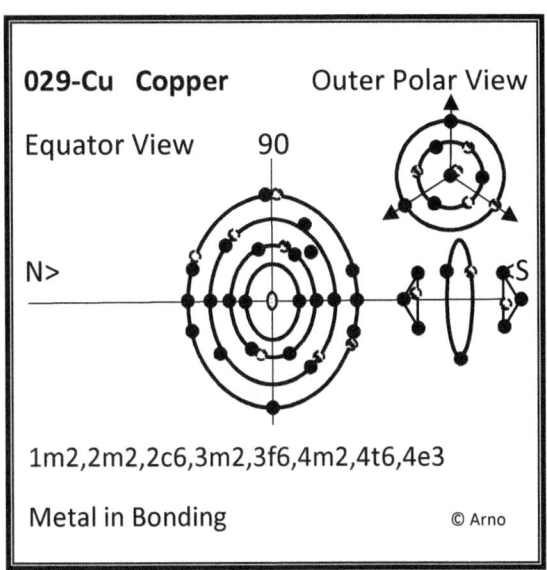

Shell/Subshell/Count	# Electrons	Structure
1m2	2	Magnetic Polar
2m2	2	Magnetic Polar
2c6	6	Scrunched Cube
3m2	2	Magnetic Polar
3f6	6	Faces of Cube
4m2	2	Magnetic Polar
4t6	6	Endcap Tetrahedron
4e3	3	Equatorial

Copper has three (3) electrons in much lower binding energy (more distance from the nucleus).

For electricity conductivity, Copper has the most electrons (3) at the easiest ratio

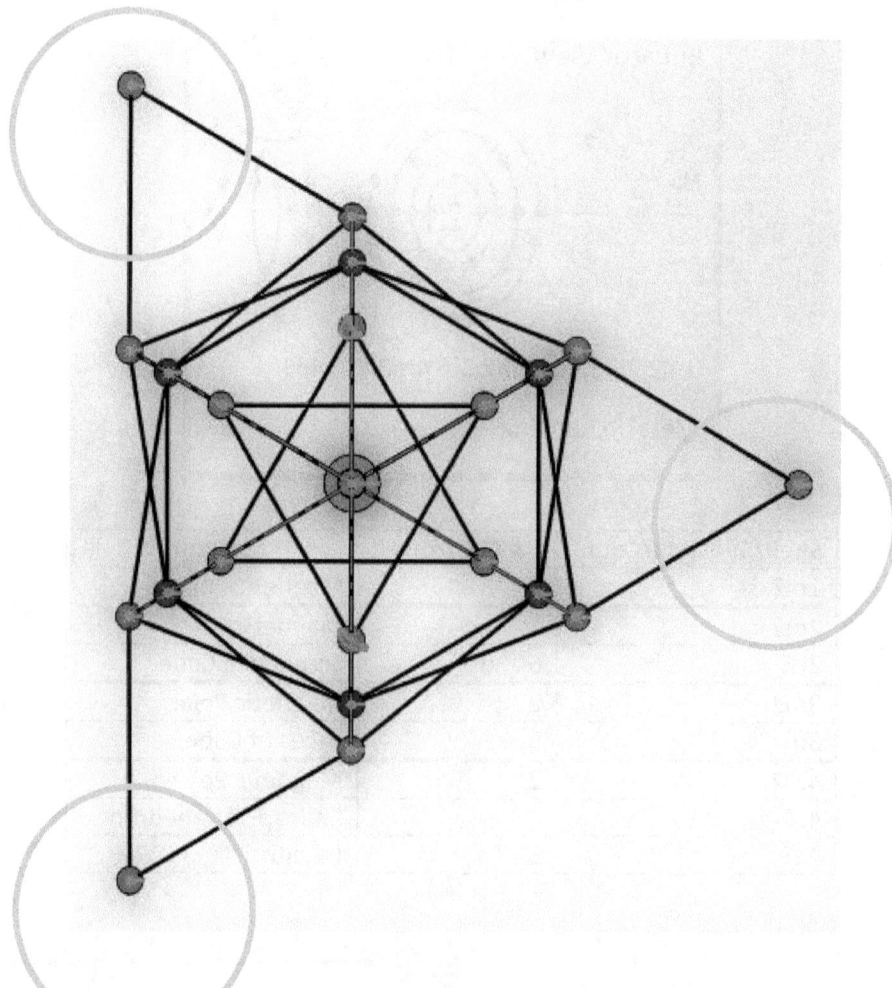

At about twice the binding distance, these 3 are 4x more likely to release. In a picture, it sticks out a lot more than inner layers.

For spectrum lines, those same three overwhelm all others to provide only a few, very, very bright electromagnetic spectrum lines.

The exchange of these three (3) 4e3 electrons creates from both less binding-energy plus the hugely large area of exposure. The few lines are very, very bright.

Copper

Compare this to Iron with many equally exposed electrons.

Even though they have the similar number of electrons, the Copper spectrum reflects the special structure of 4e3.

030-Zn Zinc

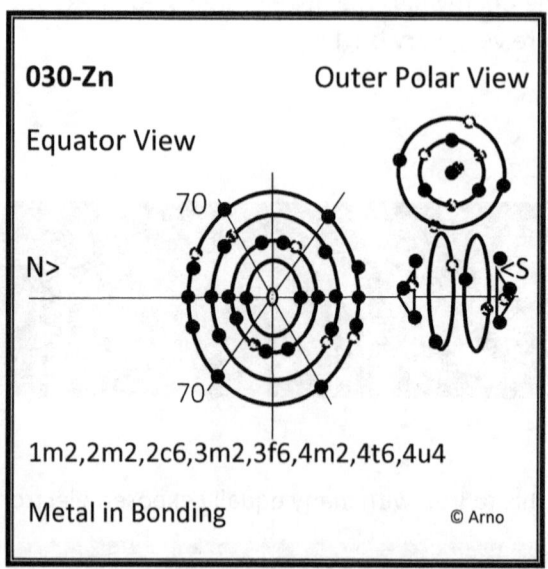

Shell/Subshell/Count	# Electrons	Structure
1m2	2	Magnetic Polar
2m2	2	Magnetic Polar
2c6	6	Scrunched Cube
3m2	2	Magnetic Polar
3f6	6	Faces of Cube
4m2	2	Magnetic Polar
4t6	6	Endcap Tetrahedron
4u4	4	Upper Longitudinal

In Zinc, the equatorial, transitional 4e3 subshells does not appear. Instead, all four electrons fill the full configuration 4u4 upper longitude subshell.

031-Ga   Gallium

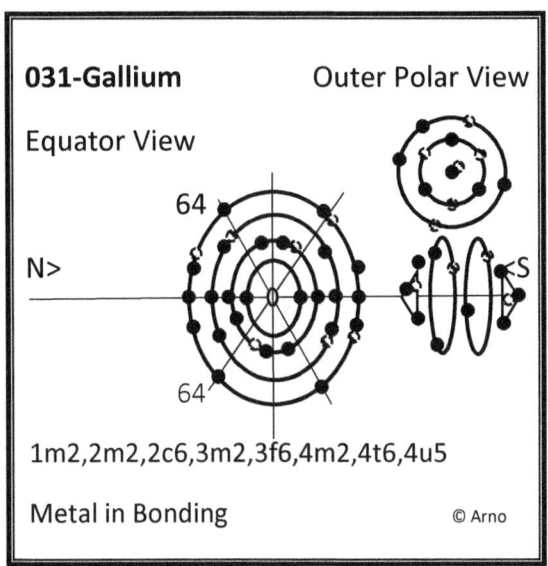

Shell/Subshell/Count	# Electrons	Structure
1m2	2	Magnetic Polar
2m2	2	Magnetic Polar
2c6	6	Scrunched Cube
3m2	2	Magnetic Polar
3f6	6	Faces of Cube
4m2	2	Magnetic Polar
4t6	6	Endcap Tetrahedron
4u5	5	Upper Longitudinal

032-Ge Germanium

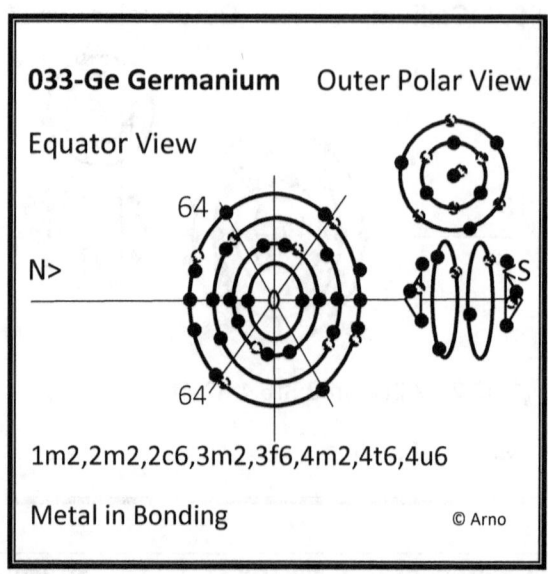

Shell/Subshell/Count	# Electrons	Structure
1m2	2	Magnetic Polar
2m2	2	Magnetic Polar
2c6	6	Scrunched Cube
3m2	2	Magnetic Polar
3f6	6	Faces of Cube
4m2	2	Magnetic Polar
4t6	6	Endcap Tetrahedron
4u6	6	Upper Longitudinal

033-As Arsenic

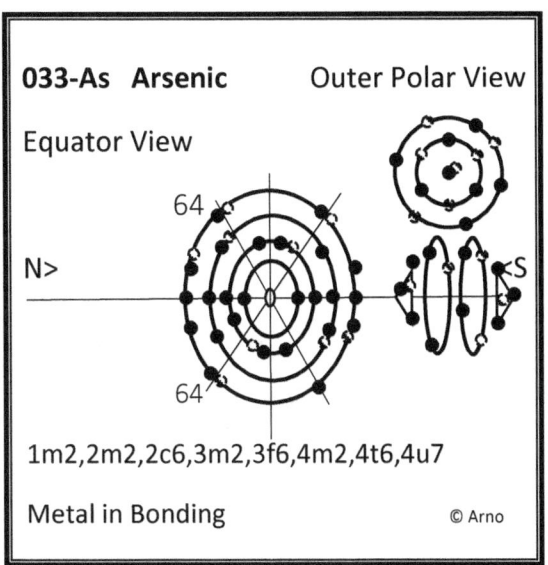

Shell/Subshell/Count	# Electrons	Structure
1m2	2	Magnetic Polar
2m2	2	Magnetic Polar
2c6	6	Scrunched Cube
3m2	2	Magnetic Polar
3f6	6	Faces of Cube
4m2	2	Magnetic Polar
4t6	6	Endcap Tetrahedron
4u7	7	Upper Longitudinal

034-Se Selenium

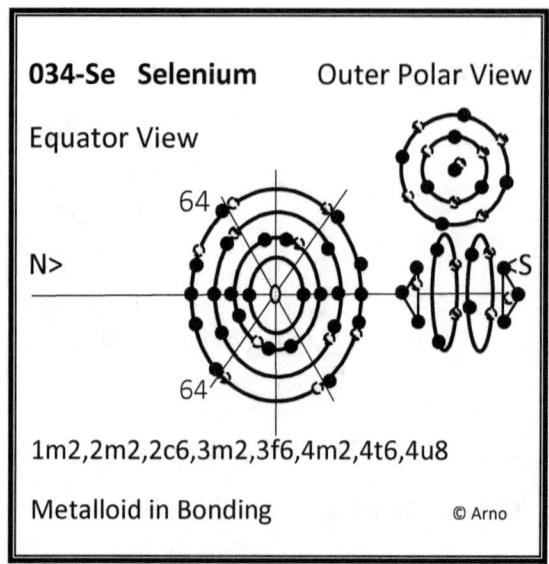

Shell/Subshell/Count	# Electrons	Structure
1m2	2	Magnetic Polar
2m2	2	Magnetic Polar
2c6	6	Scrunched Cube
3m2	2	Magnetic Polar
3f6	6	Faces of Cube
4m2	2	Magnetic Polar
4t6	6	Endcap Tetrahedron
4u8	8	Upper Longitudinal

035-Br Bromine

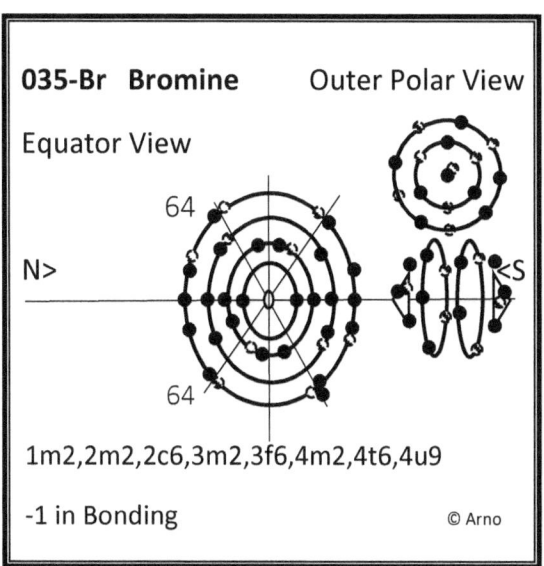

Shell/Subshell/Count	# Electrons	Structure
1m2	2	Magnetic Polar
2m2	2	Magnetic Polar
2c6	6	Scrunched Cube
3m2	2	Magnetic Polar
3f6	6	Faces of Cube
4m2	2	Magnetic Polar
4t6	6	Endcap Tetrahedron
4u9	9	Upper Longitudinal

036-Kr Krypton

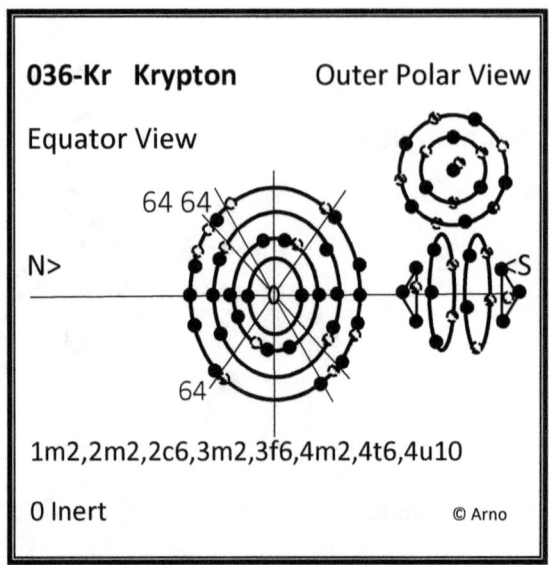

Shell/Subshell/Count	# Electrons	Structure
1m2	2	Magnetic Polar
2m2	2	Magnetic Polar
2c6	6	Scrunched Cube
3m2	2	Magnetic Polar
3f6	6	Faces of Cube
4m2	2	Magnetic Polar
4t6	6	Endcap Tetrahedron
4u10	10	Upper Longitudinal

037-Rb Rubidium

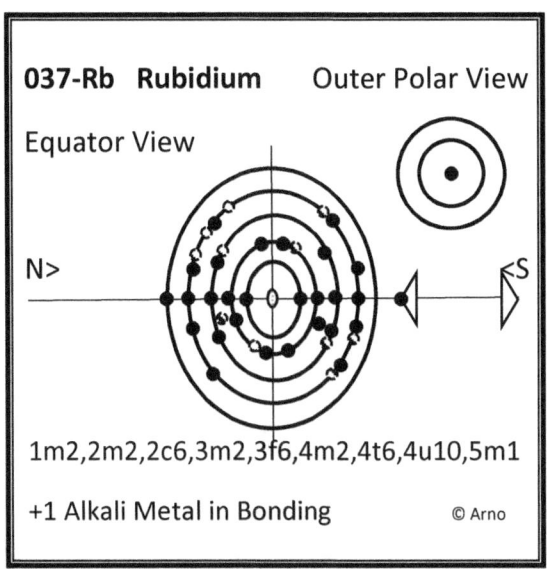

Shell/Subshell/Count	# Electrons	Structure
1m2	2	Magnetic Polar
2m2	2	Magnetic Polar
2c6	6	Scrunched Cube
3m2	2	Magnetic Polar
3f6	6	Faces of Cube
4m2	2	Magnetic Polar
4t6	6	Endcap Tetrahedron
4u10	10	Upper Longitudinal
5m1	1	Magnetic Polar

038-Sr Strontium

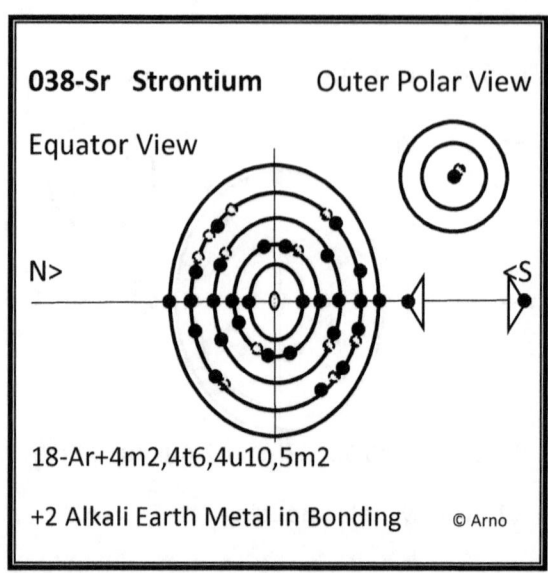

Shell/Subshell/Count	# Electrons	Structure
1m2	2	Magnetic Polar
2m2	2	Magnetic Polar
2c6	6	Scrunched Cube
3m2	2	Magnetic Polar
3f6	6	Faces of Cube
4m2	2	Magnetic Polar
4t6	6	Endcap Tetrahedron
4u10	10	Upper Longitudinal
5m2	2	Magnetic Polar

039-Y Yttrium

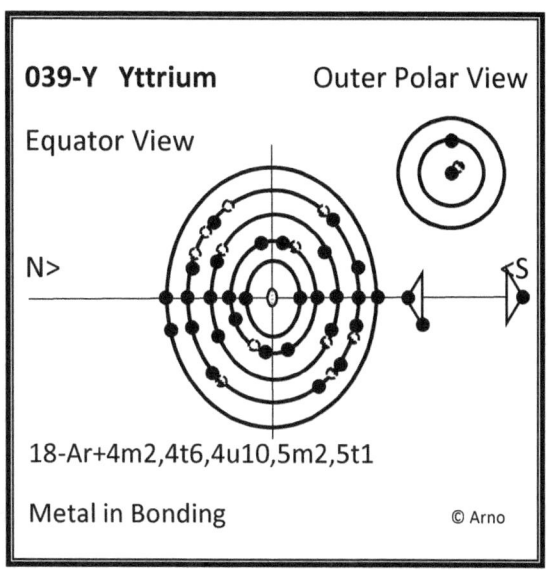

Shell/Subshell/Count	# Electrons	Structure
1m2	2	Magnetic Polar
2m2	2	Magnetic Polar
2c6	6	Scrunched Cube
3m2	2	Magnetic Polar
3f6	6	Faces of Cube
4m2	2	Magnetic Polar
4t6	6	Endcap Tetrahedron
4u10	10	Upper Longitudinal
5m2	2	Magnetic Polar
5t1	1	Endcap Tetrahedron

040-Zr Zirconium

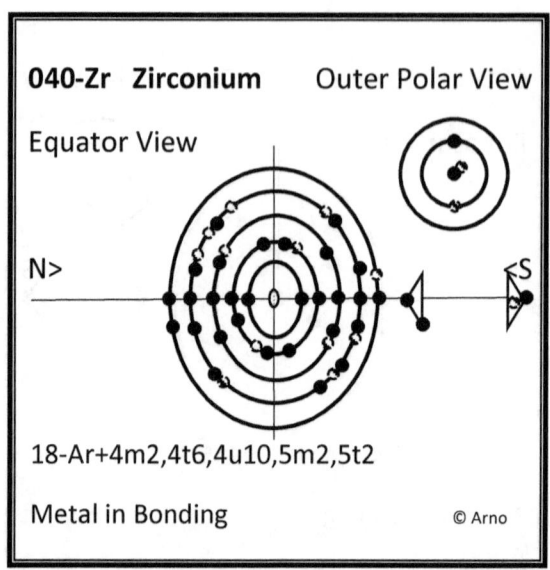

Shell/Subshell/Count	# Electrons	Structure
1m2	2	Magnetic Polar
2m2	2	Magnetic Polar
2c6	6	Scrunched Cube
3m2	2	Magnetic Polar
3f6	6	Faces of Cube
4m2	2	Magnetic Polar
4t6	6	Endcap Tetrahedron
4u10	10	Upper Longitudinal
5m2	2	Magnetic Polar
5t2	2	Endcap Tetrahedron

041-Nb Niobium

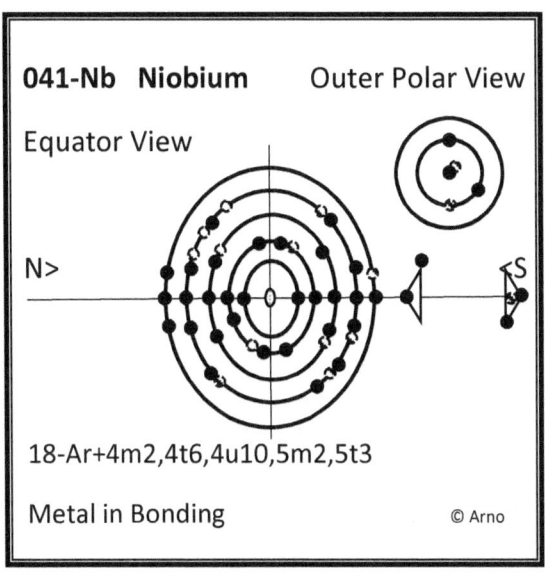

Shell/Subshell/Count	# Electrons	Structure
1m2	2	Magnetic Polar
2m2	2	Magnetic Polar
2c6	6	Scrunched Cube
3m2	2	Magnetic Polar
3f6	6	Faces of Cube
4m2	2	Magnetic Polar
4t6	6	Endcap Tetrahedron
4u10	10	Upper Longitudinal
5m2	2	Magnetic Polar
5t3	3	Endcap Tetrahedron

042-Mo Molybdenum

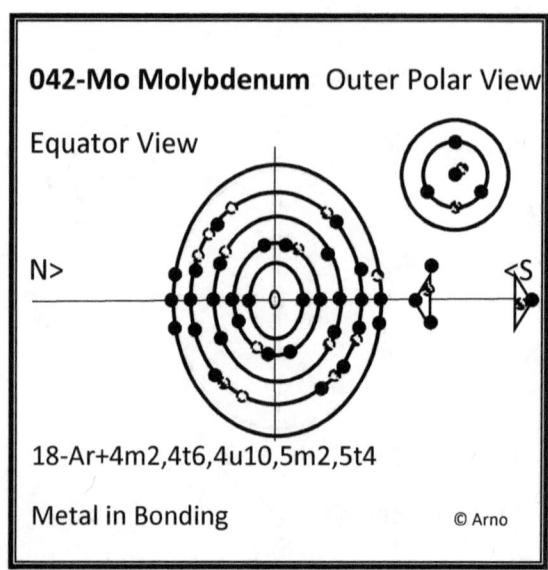

Shell/Subshell/Count	# Electrons	Structure
1m2	2	Magnetic Polar
2m2	2	Magnetic Polar
2c6	6	Scrunched Cube
3m2	2	Magnetic Polar
3f6	6	Faces of Cube
4m2	2	Magnetic Polar
4t6	6	Endcap Tetrahedron
4u10	10	Upper Longitudinal
5m2	2	Magnetic Polar
5t4	4	Endcap Tetrahedron

043-Tc Technetium

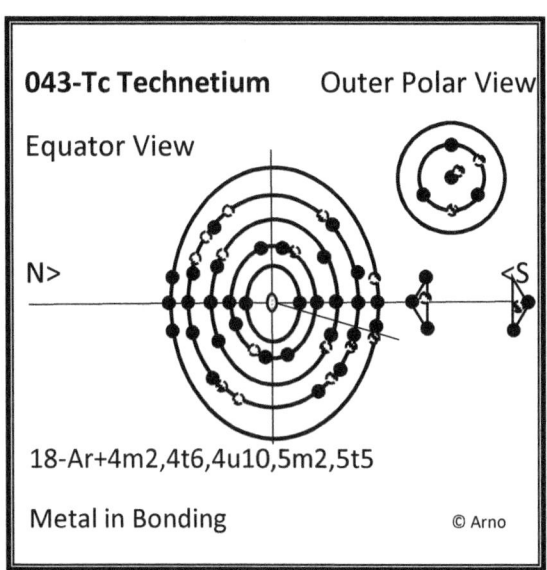

Shell/Subshell/Count	# Electrons	Structure
1m2	2	Magnetic Polar
2m2	2	Magnetic Polar
2c6	6	Scrunched Cube
3m2	2	Magnetic Polar
3f6	6	Faces of Cube
4m2	2	Magnetic Polar
4t6	6	Endcap Tetrahedron
4u10	10	Upper Longitudinal
5m2	2	Magnetic Polar
5t5	5	Endcap Tetrahedron

044-Ru Ruthenium

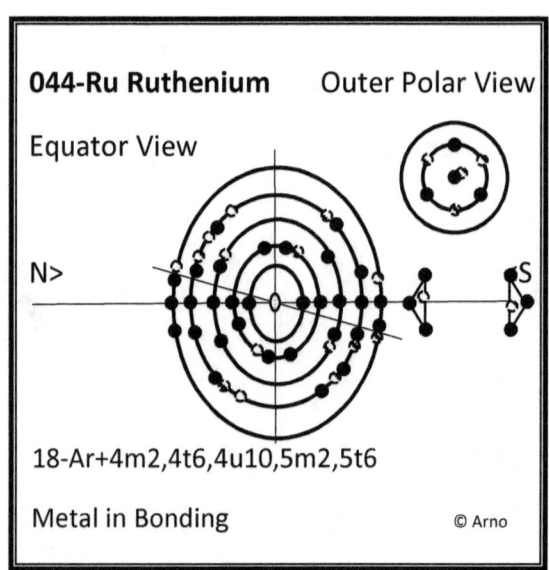

Shell/Subshell/Count	# Electrons	Structure
1m2	2	Magnetic Polar
2m2	2	Magnetic Polar
2c6	6	Scrunched Cube
3m2	2	Magnetic Polar
3f6	6	Faces of Cube
4m2	2	Magnetic Polar
4t6	6	Endcap Tetrahedron
4u10	10	Upper Longitudinal
5m2	2	Magnetic Polar
5t6	6	Endcap Tetrahedron

045-Rh Rhodium

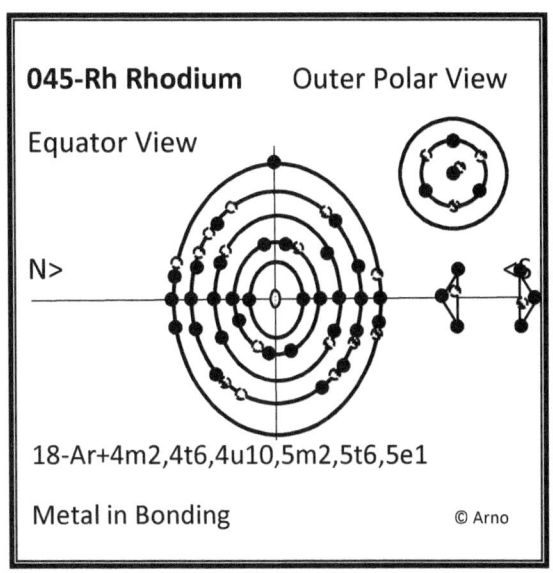

Shell/Subshell/Count	# Electrons	Structure
1m2	2	Magnetic Polar
2m2	2	Magnetic Polar
2c6	6	Scrunched Cube
3m2	2	Magnetic Polar
3f6	6	Faces of Cube
4m2	2	Magnetic Polar
4t6	6	Endcap Tetrahedron
4u10	10	Upper Longitudinal
5m2	2	Magnetic Polar
5t6	6	Endcap Tetrahedron
5e1	1	Equatorial

046-Pd Palladium

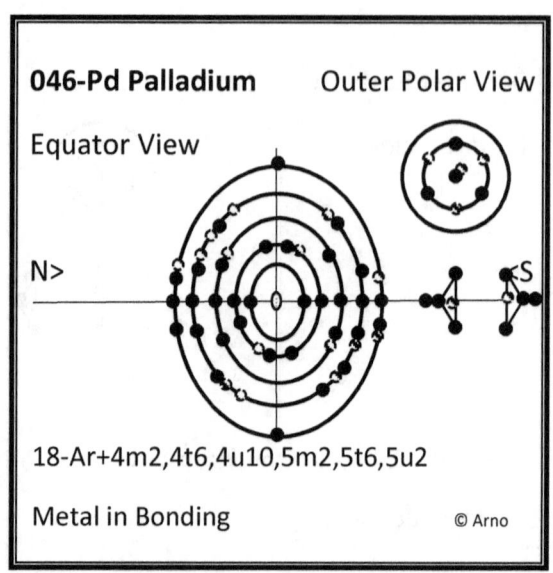

Shell/Subshell/Count	# Electrons	Structure
1m2	2	Magnetic Polar
2m2	2	Magnetic Polar
2c6	6	Scrunched Cube
3m2	2	Magnetic Polar
3f6	6	Faces of Cube
4m2	2	Magnetic Polar
4t6	6	Endcap Tetrahedron
4u10	10	Upper Longitudinal
5m2	2	Magnetic Polar
5t6	6	Endcap Tetrahedron
5e2	2	Equatorial

047-Ag Silver

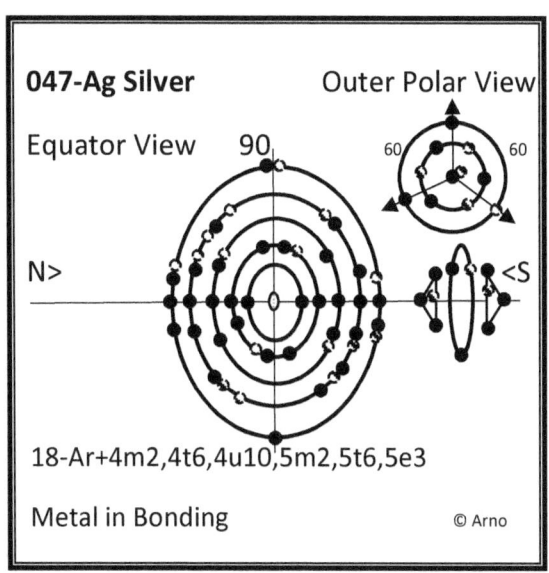

Shell/Subshell/Count	# Electrons	Structure
1m2	2	Magnetic Polar
2m2	2	Magnetic Polar
2c6	6	Scrunched Cube
3m2	2	Magnetic Polar
3f6	6	Faces of Cube
4m2	2	Magnetic Polar
4t6	6	Endcap Tetrahedron
4u10	10	Upper Longitudinal
5m2	2	Magnetic Polar
5t6	6	Endcap Tetrahedron
5e3	3	Equatorial

048-Cd Cadmium

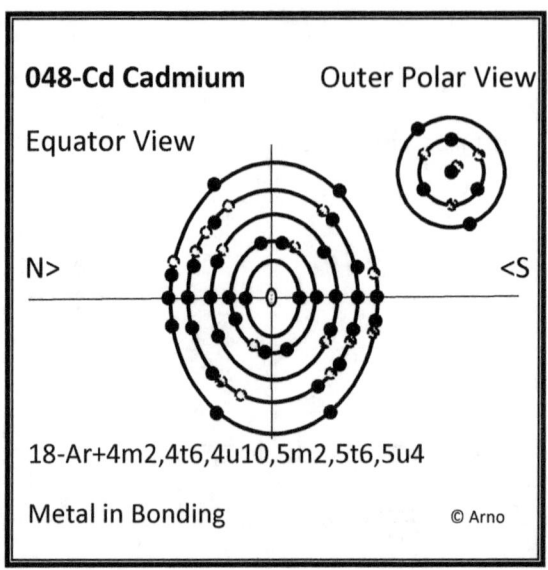

Shell/Subshell/Count	# Electrons	Structure
1m2	2	Magnetic Polar
2m2	2	Magnetic Polar
2c6	6	Scrunched Cube
3m2	2	Magnetic Polar
3f6	6	Faces of Cube
4m2	2	Magnetic Polar
4t6	6	Endcap Tetrahedron
4u10	10	Upper Longitudinal
5m2	2	Magnetic Polar
5t6	6	Endcap Tetrahedron
5u4	4	Upper Longitudinal

049-In Indium

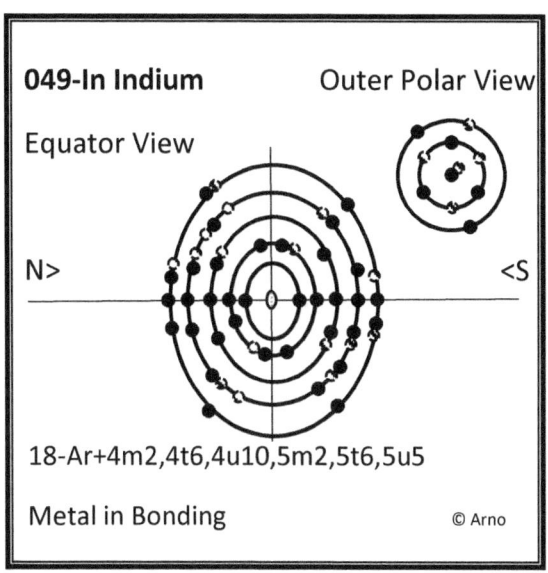

Shell/Subshell/Count	# Electrons	Structure
1m2	2	Magnetic Polar
2m2	2	Magnetic Polar
2c6	6	Scrunched Cube
3m2	2	Magnetic Polar
3f6	6	Faces of Cube
4m2	2	Magnetic Polar
4t6	6	Endcap Tetrahedron
4u10	10	Upper Longitudinal
5m2	2	Magnetic Polar
5t6	6	Endcap Tetrahedron
5u5	5	Upper Longitudinal

050-Sn Tin

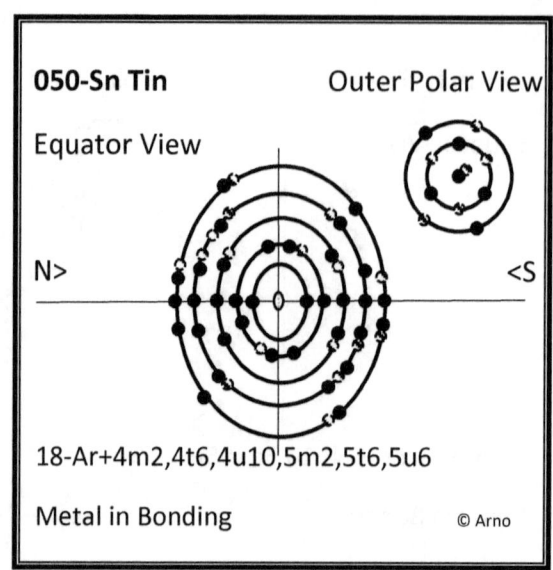

Shell/Subshell/Count	# Electrons	Structure
1m2	2	Magnetic Polar
2m2	2	Magnetic Polar
2c6	6	Scrunched Cube
3m2	2	Magnetic Polar
3f6	6	Faces of Cube
4m2	2	Magnetic Polar
4t6	6	Endcap Tetrahedron
4u10	10	Upper Longitudinal
5m2	2	Magnetic Polar
5t6	6	Endcap Tetrahedron
5u6	6	Upper Longitudinal

051-Sb Antimony

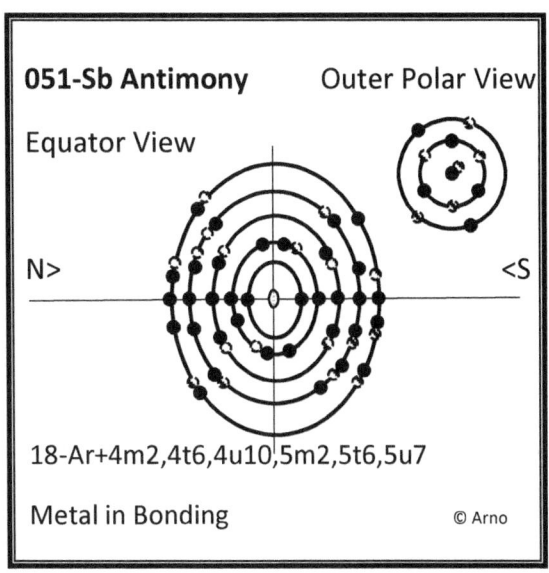

Shell/Subshell/Count	# Electrons	Structure
1m2	2	Magnetic Polar
2m2	2	Magnetic Polar
2c6	6	Scrunched Cube
3m2	2	Magnetic Polar
3f6	6	Faces of Cube
4m2	2	Magnetic Polar
4t6	6	Endcap Tetrahedron
4u10	10	Upper Longitudinal
5m2	2	Magnetic Polar
5t6	6	Endcap Tetrahedron
5u7	7	Upper Longitudinal

052-Te Tellurium

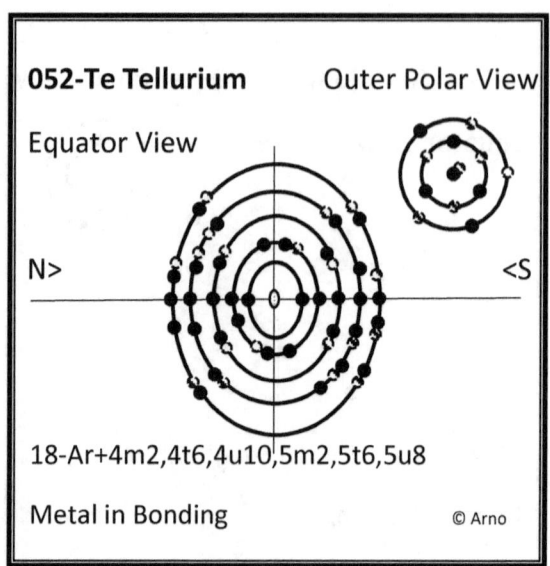

Shell/Subshell/Count	# Electrons	Structure
1m2	2	Magnetic Polar
2m2	2	Magnetic Polar
2c6	6	Scrunched Cube
3m2	2	Magnetic Polar
3f6	6	Faces of Cube
4m2	2	Magnetic Polar
4t6	6	Endcap Tetrahedron
4u10	10	Upper Longitudinal
5m2	2	Magnetic Polar
5t6	6	Endcap Tetrahedron
5u8	8	Upper Longitudinal

053-I Iodine

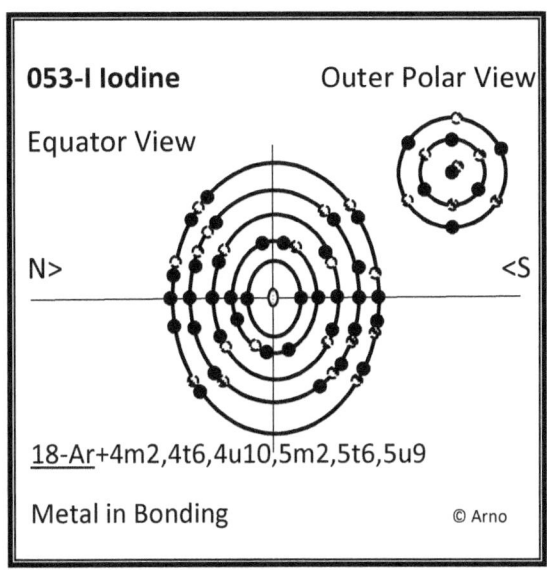

Shell/Subshell/Count	# Electrons	Structure
1m2	2	Magnetic Polar
2m2	2	Magnetic Polar
2c6	6	Scrunched Cube
3m2	2	Magnetic Polar
3f6	6	Faces of Cube
4m2	2	Magnetic Polar
4t6	6	Endcap Tetrahedron
4u10	10	Upper Longitudinal
5m2	2	Magnetic Polar
5t6	6	Endcap Tetrahedron
5u9	9	Upper Longitudinal

054-Xe Xenon

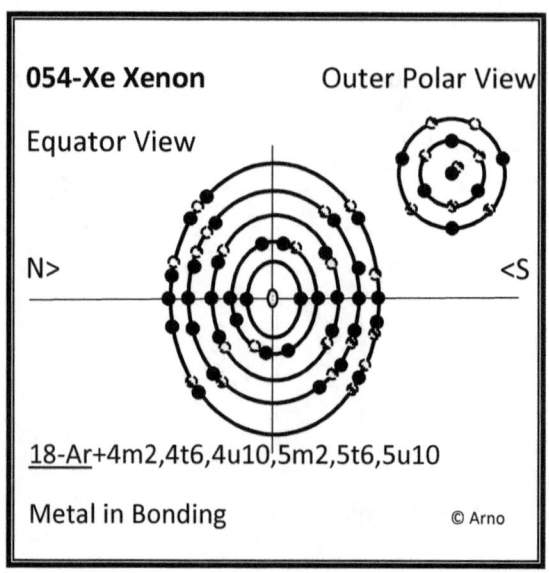

Shell/Subshell/Count	# Electrons	Structure
1m2	2	Magnetic Polar
2m2	2	Magnetic Polar
2c6	6	Scrunched Cube
3m2	2	Magnetic Polar
3f6	6	Faces of Cube
4m2	2	Magnetic Polar
4t6	6	Endcap Tetrahedron
4u10	10	Upper Longitudinal
5m2	2	Magnetic Polar
5t6	6	Endcap Tetrahedron
5u10	10	Upper Longitudinal

055-Cs Cesium/Caesium

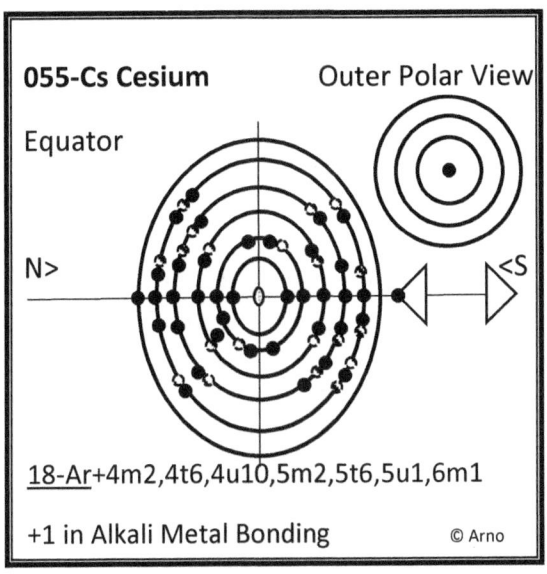

Shell/Subshell/Count	# Electrons	Structure
1m2	2	Magnetic Polar
2m2	2	Magnetic Polar
2c6	6	Scrunched Cube
3m2	2	Magnetic Polar
3f6	6	Faces of Cube
4m2	2	Magnetic Polar
4t6	6	Endcap Tetrahedron
4u10	10	Upper Longitudinal
5m2	2	Magnetic Polar
5t6	6	Endcap Tetrahedron
5u10	10	Upper Longitudinal
6m1	1	Magnetic Polar

056-Ba Barium

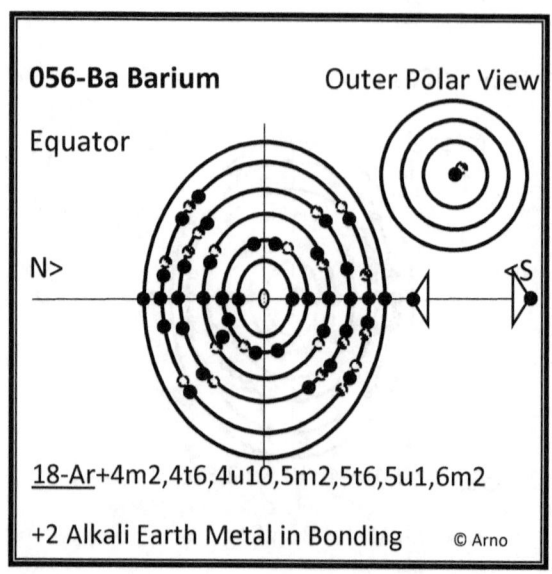

Shell/Subshell/Count	# Electrons	Structure
1m2	2	Magnetic Polar
2m2	2	Magnetic Polar
2c6	6	Scrunched Cube
3m2	2	Magnetic Polar
3f6	6	Faces of Cube
4m2	2	Magnetic Polar
4t6	6	Endcap Tetrahedron
4u10	10	Upper Longitudinal
5m2	2	Magnetic Polar
5t6	6	Endcap Tetrahedron
5u10	10	Upper Longitudinal
6m2	2	Magnetic Polar

057-La Lanthanum

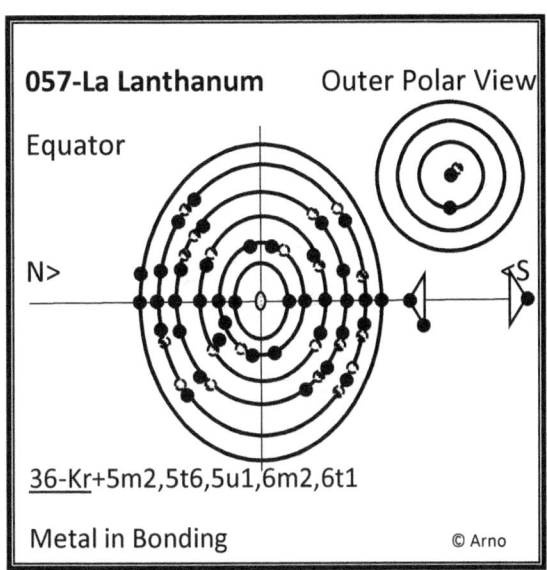

Shell/Subshell/Count	# Electrons	Structure
1m2	2	Magnetic Polar
2m2	2	Magnetic Polar
2c6	6	Scrunched Cube
3m2	2	Magnetic Polar
3f6	6	Faces of Cube
4m2	2	Magnetic Polar
4t6	6	Endcap Tetrahedron
4u10	10	Upper Longitudinal
5m2	2	Magnetic Polar
5t6	6	Endcap Tetrahedron
5u10	10	Upper Longitudinal
6m2	2	Magnetic Polar
6t1	1	Endcap Tetrahedron

058-Ce Cerium

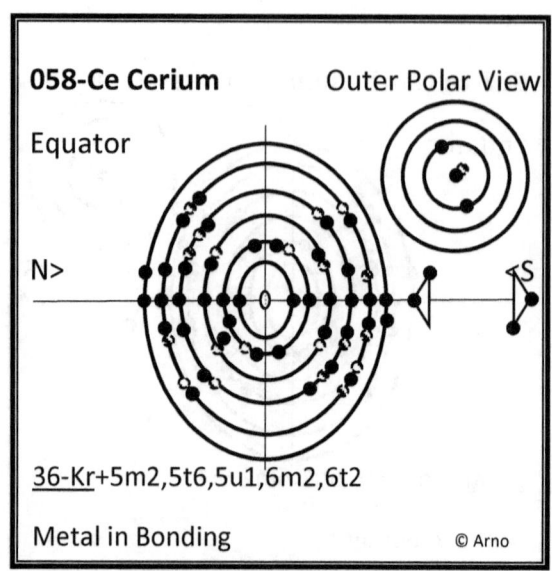

Shell/Subshell/Count	# Electrons	Structure
1m2	2	Magnetic Polar
2m2	2	Magnetic Polar
2c6	6	Scrunched Cube
3m2	2	Magnetic Polar
3f6	6	Faces of Cube
4m2	2	Magnetic Polar
4t6	6	Endcap Tetrahedron
4u10	10	Upper Longitudinal
5m2	2	Magnetic Polar
5t6	6	Endcap Tetrahedron
5u10	10	Upper Longitudinal
6m2	2	Magnetic Polar
6t2	2	Endcap Tetrahedron

059-Pr Praseodymium

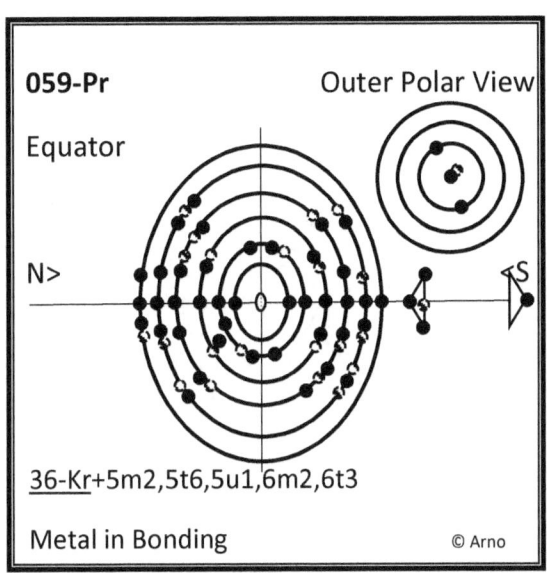

Shell/Subshell/Count	# Electrons	Structure
1m2	2	Magnetic Polar
2m2	2	Magnetic Polar
2c6	6	Scrunched Cube
3m2	2	Magnetic Polar
3f6	6	Faces of Cube
4m2	2	Magnetic Polar
4t6	6	Endcap Tetrahedron
4u10	10	Upper Longitudinal
5m2	2	Magnetic Polar
5t6	6	Endcap Tetrahedron
5u10	10	Upper Longitudinal
6m2	2	Magnetic Polar
6t3	3	Endcap Tetrahedron

060-Nd Neodymium

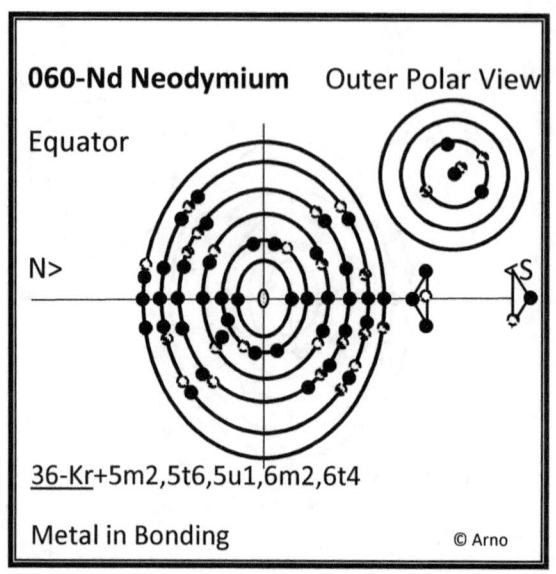

Shell/Subshell/Count	# Electrons	Structure
1m2	2	Magnetic Polar
2m2	2	Magnetic Polar
2c6	6	Scrunched Cube
3m2	2	Magnetic Polar
3f6	6	Faces of Cube
4m2	2	Magnetic Polar
4t6	6	Endcap Tetrahedron
4u10	10	Upper Longitudinal
5m2	2	Magnetic Polar
5t6	6	Endcap Tetrahedron
5u10	10	Upper Longitudinal
6m2	2	Magnetic Polar
6t4	4	Endcap Tetrahedron

061-Pm Promethium

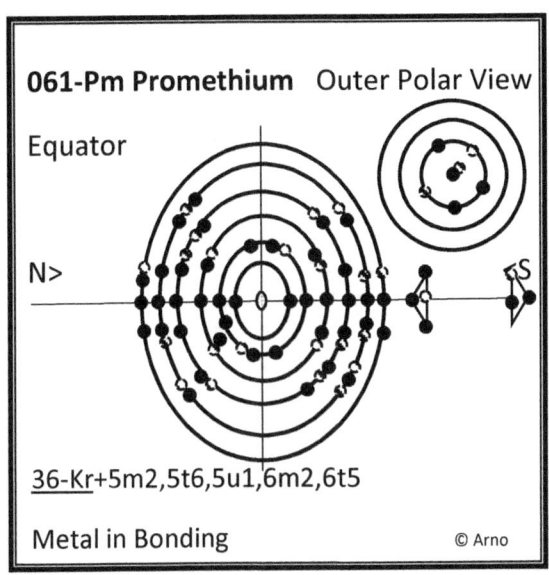

Shell/Subshell/Count	# Electrons	Structure
1m2	2	Magnetic Polar
2m2	2	Magnetic Polar
2c6	6	Scrunched Cube
3m2	2	Magnetic Polar
3f6	6	Faces of Cube
4m2	2	Magnetic Polar
4t6	6	Endcap Tetrahedron
4u10	10	Upper Longitudinal
5m2	2	Magnetic Polar
5t6	6	Endcap Tetrahedron
5u10	10	Upper Longitudinal
6m2	2	Magnetic Polar
6t5	5	Endcap Tetrahedron

062-Sm Samarium

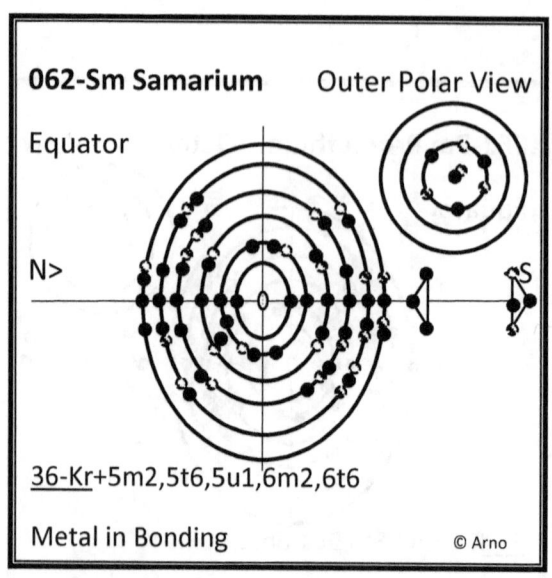

Shell/Subshell/Count	# Electrons	Structure
1m2	2	Magnetic Polar
2m2	2	Magnetic Polar
2c6	6	Scrunched Cube
3m2	2	Magnetic Polar
3f6	6	Faces of Cube
4m2	2	Magnetic Polar
4t6	6	Endcap Tetrahedron
4u10	10	Upper Longitudinal
5m2	2	Magnetic Polar
5t6	6	Endcap Tetrahedron
5u10	10	Upper Longitudinal
6m2	2	Magnetic Polar
6t6	6	Endcap Tetrahedron

063-Eu Europium

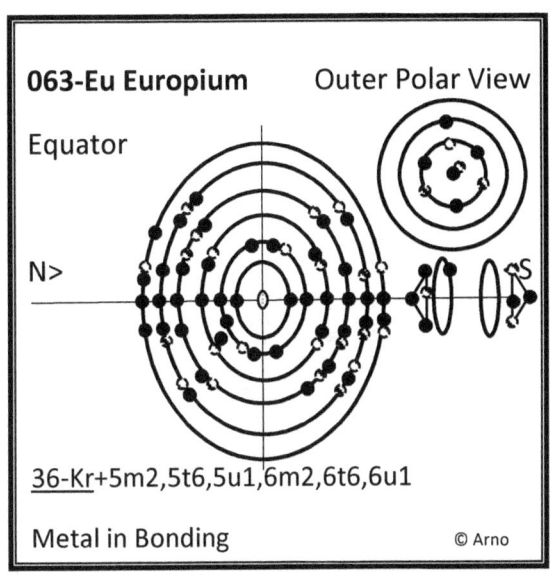

Shell/Subshell/Count	# Electrons	Structure
1m2	2	Magnetic Polar
2m2	2	Magnetic Polar
2c6	6	Scrunched Cube
3m2	2	Magnetic Polar
3f6	6	Faces of Cube
4m2	2	Magnetic Polar
4t6	6	Endcap Tetrahedron
4u10	10	Upper Longitudinal
5m2	2	Magnetic Polar
5t6	6	Endcap Tetrahedron
5u10	10	Upper Longitudinal
6m2	2	Magnetic Polar
6t6	6	Endcap Tetrahedron
6u1	1	Upper Longitudinal

064-Gd Gadolinium

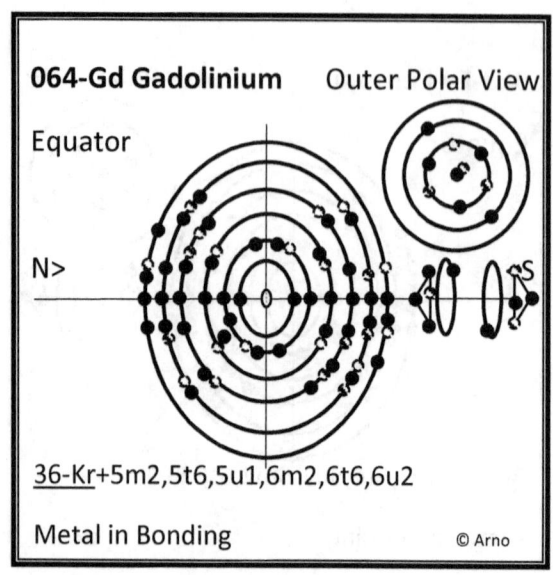

Shell/Subshell/Count	# Electrons	Structure
1m2	2	Magnetic Polar
2m2	2	Magnetic Polar
2c6	6	Scrunched Cube
3m2	2	Magnetic Polar
3f6	6	Faces of Cube
4m2	2	Magnetic Polar
4t6	6	Endcap Tetrahedron
4u10	10	Upper Longitudinal
5m2	2	Magnetic Polar
5t6	6	Endcap Tetrahedron
5u10	10	Upper Longitudinal
6m2	2	Magnetic Polar
6t6	6	Endcap Tetrahedron
6u2	2	Upper Longitudinal

065-Tb Terbium

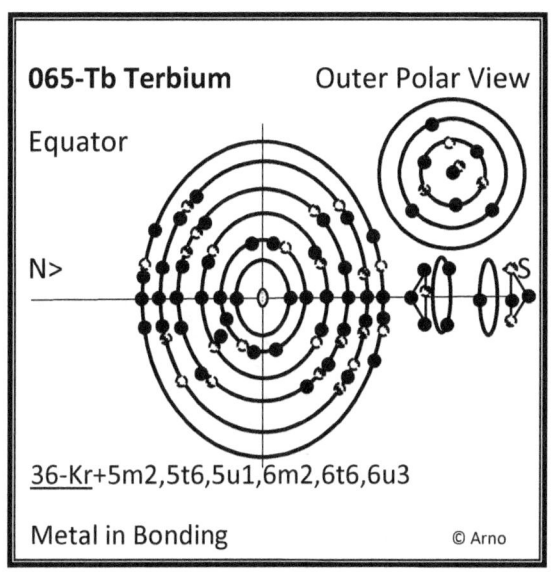

Shell/Subshell/Count	# Electrons	Structure
1m2	2	Magnetic Polar
2m2	2	Magnetic Polar
2c6	6	Scrunched Cube
3m2	2	Magnetic Polar
3f6	6	Faces of Cube
4m2	2	Magnetic Polar
4t6	6	Endcap Tetrahedron
4u10	10	Upper Longitudinal
5m2	2	Magnetic Polar
5t6	6	Endcap Tetrahedron
5u10	10	Upper Longitudinal
6m2	2	Magnetic Polar
6t6	6	Endcap Tetrahedron
6u3	3	Upper Longitudinal

066-Dy Dysprosium

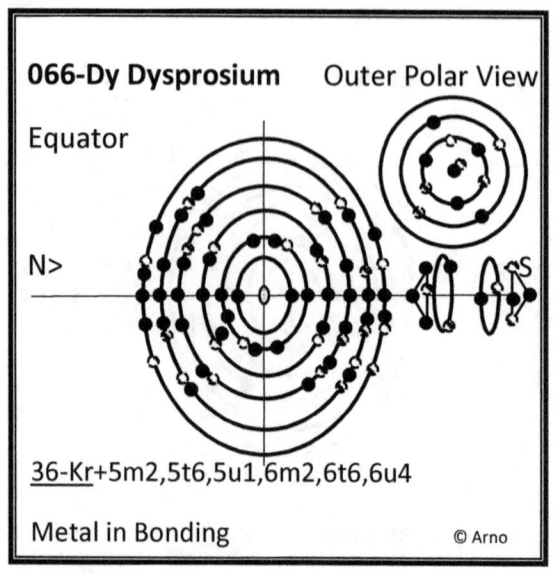

Shell/Subshell/Count	# Electrons	Structure
1m2	2	Magnetic Polar
2m2	2	Magnetic Polar
2c6	6	Scrunched Cube
3m2	2	Magnetic Polar
3f6	6	Faces of Cube
4m2	2	Magnetic Polar
4t6	6	Endcap Tetrahedron
4u10	10	Upper Longitudinal
5m2	2	Magnetic Polar
5t6	6	Endcap Tetrahedron
5u10	10	Upper Longitudinal
6m2	2	Magnetic Polar
6t6	6	Endcap Tetrahedron
6u4	4	Upper Longitudinal

067-Ho Holmium

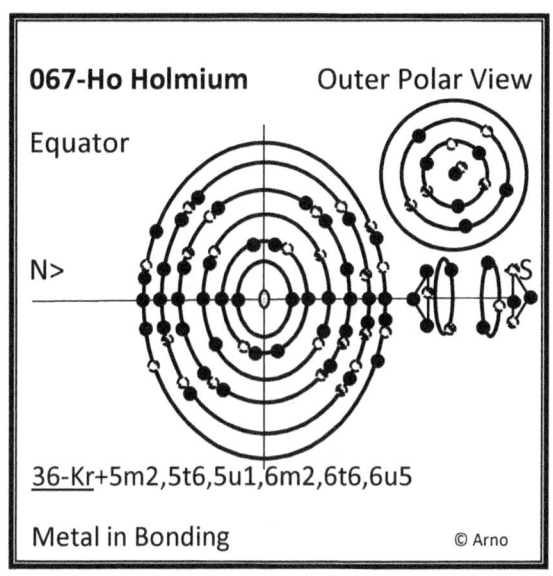

Shell/Subshell/Count	# Electrons	Structure
1m2	2	Magnetic Polar
2m2	2	Magnetic Polar
2c6	6	Scrunched Cube
3m2	2	Magnetic Polar
3f6	6	Faces of Cube
4m2	2	Magnetic Polar
4t6	6	Endcap Tetrahedron
4u10	10	Upper Longitudinal
5m2	2	Magnetic Polar
5t6	6	Endcap Tetrahedron
5u10	10	Upper Longitudinal
6m2	2	Magnetic Polar
6t6	6	Endcap Tetrahedron
6u5	5	Upper Longitudinal

068-Er Erbium

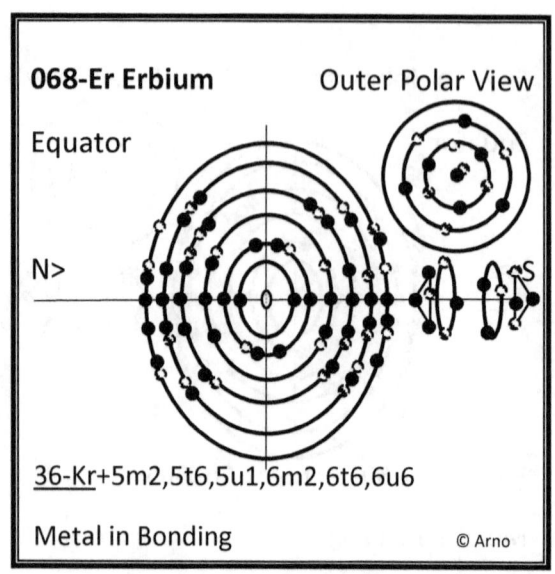

Shell/Subshell/Count	# Electrons	Structure
1m2	2	Magnetic Polar
2m2	2	Magnetic Polar
2c6	6	Scrunched Cube
3m2	2	Magnetic Polar
3f6	6	Faces of Cube
4m2	2	Magnetic Polar
4t6	6	Endcap Tetrahedron
4u10	10	Upper Longitudinal
5m2	2	Magnetic Polar
5t6	6	Endcap Tetrahedron
5u10	10	Upper Longitudinal
6m2	2	Magnetic Polar
6t6	6	Endcap Tetrahedron
6u6	6	Upper Longitudinal

069-Tm Thulium

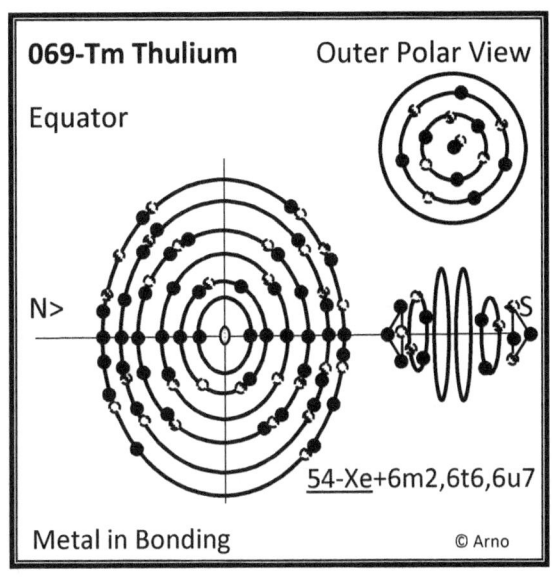

Shell/Subshell/Count	# Electrons	Structure
1m2	2	Magnetic Polar
2m2	2	Magnetic Polar
2c6	6	Scrunched Cube
3m2	2	Magnetic Polar
3f6	6	Faces of Cube
4m2	2	Magnetic Polar
4t6	6	Endcap Tetrahedron
4u10	10	Upper Longitudinal
5m2	2	Magnetic Polar
5t6	6	Endcap Tetrahedron
5u10	10	Upper Longitudinal
6m2	2	Magnetic Polar
6t6	6	Endcap Tetrahedron
6u7	7	Upper Longitudinal

070-Yb Ytterbium

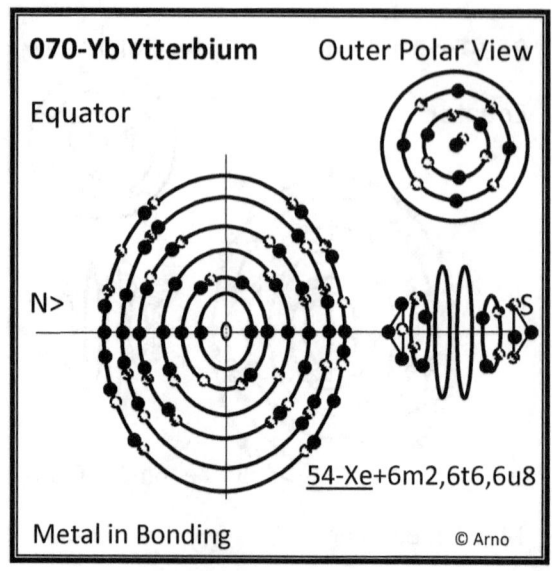

Shell/Subshell/Count	# Electrons	Structure
1m2	2	Magnetic Polar
2m2	2	Magnetic Polar
2c6	6	Scrunched Cube
3m2	2	Magnetic Polar
3f6	6	Faces of Cube
4m2	2	Magnetic Polar
4t6	6	Endcap Tetrahedron
4u10	10	Upper Longitudinal
5m2	2	Magnetic Polar
5t6	6	Endcap Tetrahedron
5u10	10	Upper Longitudinal
6m2	2	Magnetic Polar
6t6	6	Endcap Tetrahedron
6u8	8	Upper Longitudinal

071-Lu Lutetium

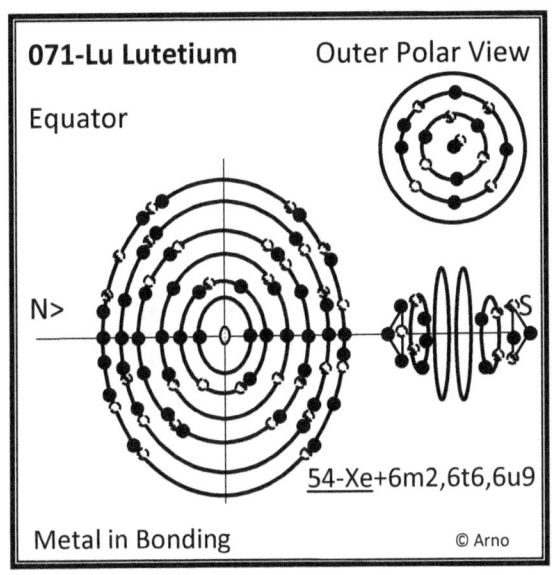

Shell/Subshell/Count	# Electrons	Structure
1m2	2	Magnetic Polar
2m2	2	Magnetic Polar
2c6	6	Scrunched Cube
3m2	2	Magnetic Polar
3f6	6	Faces of Cube
4m2	2	Magnetic Polar
4t6	6	Endcap Tetrahedron
4u10	10	Upper Longitudinal
5m2	2	Magnetic Polar
5t6	6	Endcap Tetrahedron
5u10	10	Upper Longitudinal
6m2	2	Magnetic Polar
6t6	6	Endcap Tetrahedron
6u9	9	Upper Longitudinal

072-Hf Hafnium

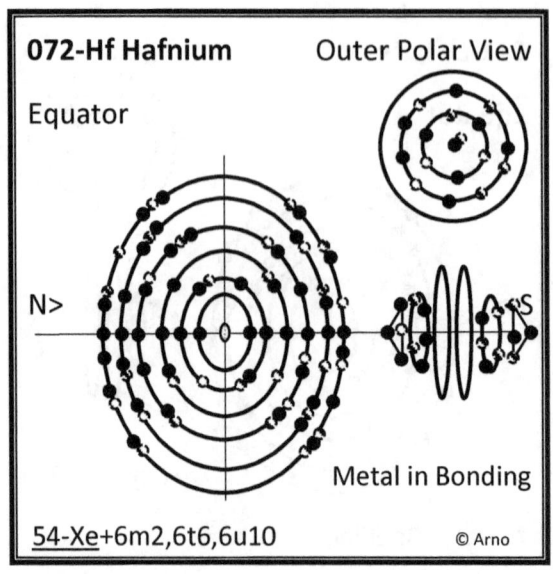

Shell/Subshell/Count	# Electrons	Structure
1m2	2	Magnetic Polar
2m2	2	Magnetic Polar
2c6	6	Scrunched Cube
3m2	2	Magnetic Polar
3f6	6	Faces of Cube
4m2	2	Magnetic Polar
4t6	6	Endcap Tetrahedron
4u10	10	Upper Longitudinal
5m2	2	Magnetic Polar
5t6	6	Endcap Tetrahedron
5u10	10	Upper Longitudinal
6m2	2	Magnetic Polar
6t6	6	Endcap Tetrahedron
6u10	10	Upper Longitudinal

073-Ta Tantalum

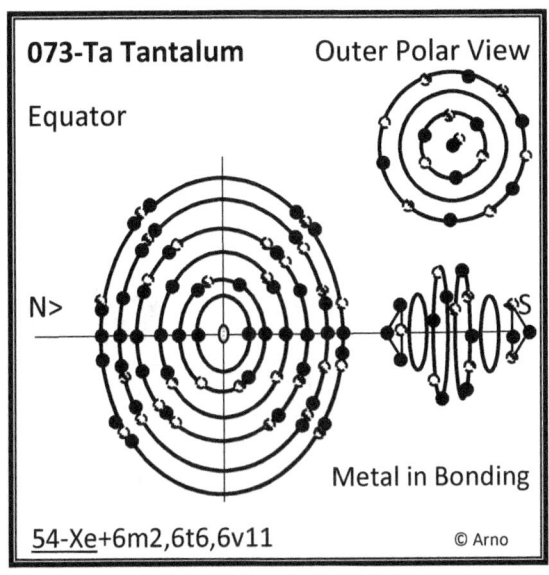

Shell/Subshell/Count	# Electrons	Structure
1m2	2	Magnetic Polar
2m2	2	Magnetic Polar
2c6	6	Scrunched Cube
3m2	2	Magnetic Polar
3f6	6	Faces of Cube
4m2	2	Magnetic Polar
4t6	6	Endcap Tetrahedron
4u10	10	Upper Longitudinal
5m2	2	Magnetic Polar
5t6	6	Endcap Tetrahedron
5u10	10	Upper Longitudinal
6m2	2	Magnetic Polar
6t6	6	Endcap Tetrahedron
6u0:10	-0-	Upper Longitudinal
6v11:14	11	Very Big Inclination

074-W Tungsten

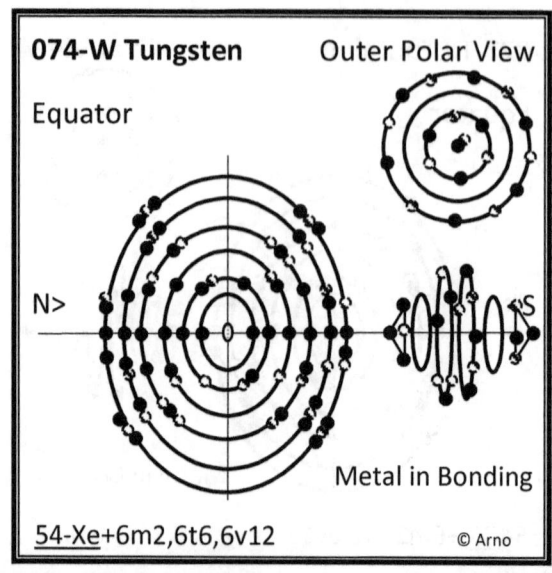

Shell/Subshell/Count	# Electrons	Structure
1m2	2	Magnetic Polar
2m2	2	Magnetic Polar
2c6	6	Scrunched Cube
3m2	2	Magnetic Polar
3f6	6	Faces of Cube
4m2	2	Magnetic Polar
4t6	6	Endcap Tetrahedron
4u10	10	Upper Longitudinal
5m2	2	Magnetic Polar
5t6	6	Endcap Tetrahedron
5u10	10	Upper Longitudinal
6m2	2	Magnetic Polar
6t6	6	Endcap Tetrahedron
6u0:10	-0-	Upper Longitudinal
6v12:14	12	Very Big Inclination

075-Re Rhenium

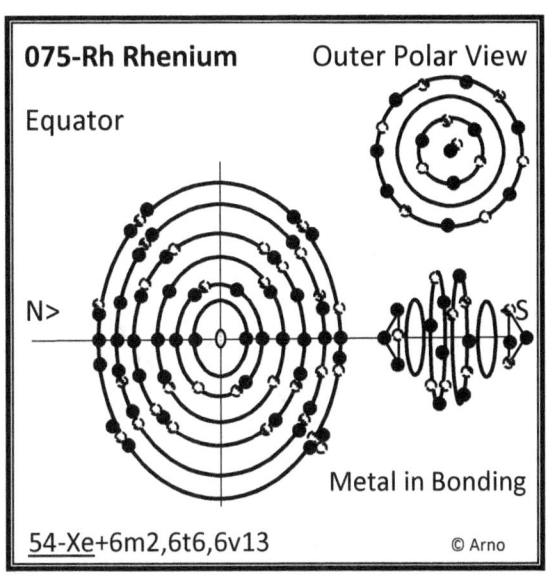

Shell/Subshell/Count	# Electrons	Structure
1m2	2	Magnetic Polar
2m2	2	Magnetic Polar
2c6	6	Scrunched Cube
3m2	2	Magnetic Polar
3f6	6	Faces of Cube
4m2	2	Magnetic Polar
4t6	6	Endcap Tetrahedron
4u10	10	Upper Longitudinal
5m2	2	Magnetic Polar
5t6	6	Endcap Tetrahedron
5u10	10	Upper Longitudinal
6m2	2	Magnetic Polar
6t6	6	Endcap Tetrahedron
6u0:10	-0-	Upper Longitudinal
6v13:14	13	Very Big Inclination

076-Os Osmium

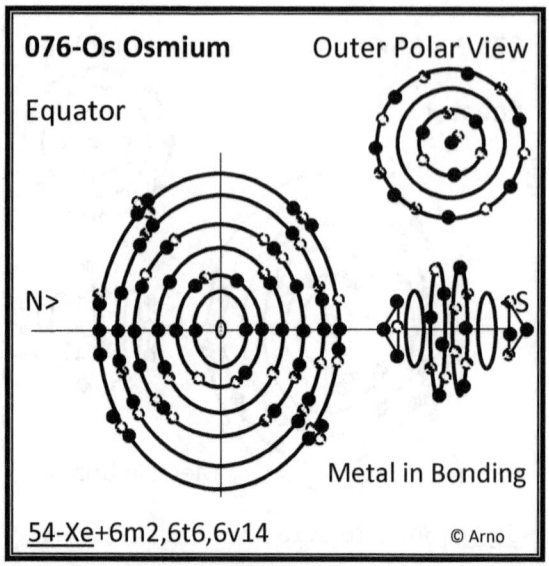

Shell/Subshell/Count	# Electrons	Structure
1m2	2	Magnetic Polar
2m2	2	Magnetic Polar
2c6	6	Scrunched Cube
3m2	2	Magnetic Polar
3f6	6	Faces of Cube
4m2	2	Magnetic Polar
4t6	6	Endcap Tetrahedron
4u10	10	Upper Longitudinal
5m2	2	Magnetic Polar
5t6	6	Endcap Tetrahedron
5u10	10	Upper Longitudinal
6m2	2	Magnetic Polar
6t6	6	Endcap Tetrahedron
6u0:14	-0-	Upper Longitudinal
6v14:14	14	Very Big Inclination

077-Ir Iridium

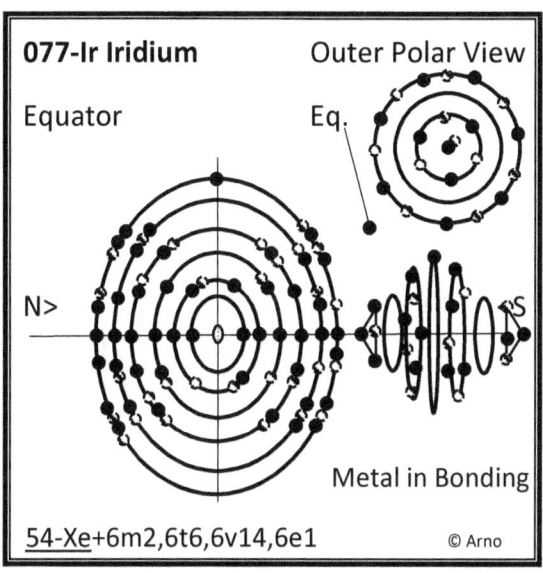

Shell/Subshell/Count	# Electrons	Structure
1m2	2	Magnetic Polar
2m2	2	Magnetic Polar
2c6	6	Scrunched Cube
3m2	2	Magnetic Polar
3f6	6	Faces of Cube
4m2	2	Magnetic Polar
4t6	6	Endcap Tetrahedron
4u10	10	Upper Longitudinal
5m2	2	Magnetic Polar
5t6	6	Endcap Tetrahedron
5u10	10	Upper Longitudinal
6m2	2	Magnetic Polar
6t6	6	Endcap Tetrahedron
6u0:10	-0-	Longitudinal
6v14:14	14	Very Big
6e1:3	1	Equatorial

078-Pt Platinum

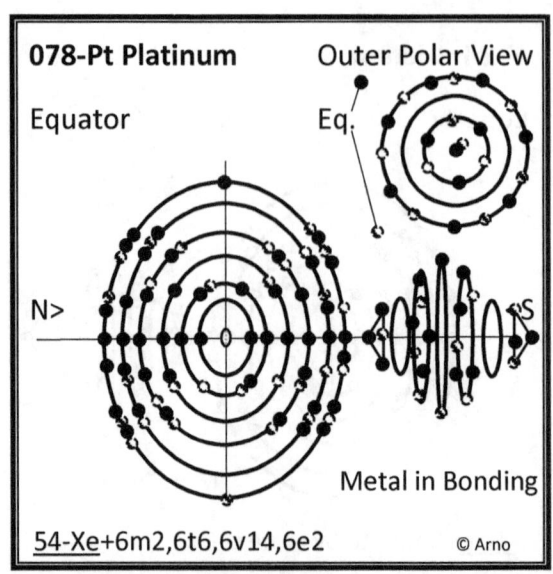

Shell/Subshell/Count	# Electrons	Structure
1m2	2	Magnetic Polar
2m2	2	Magnetic Polar
2c6	6	Scrunched Cube
3m2	2	Magnetic Polar
3f6	6	Faces of Cube
4m2	2	Magnetic Polar
4t6	6	Endcap Tetrahedron
4u10	10	Upper Longitudinal
5m2	2	Magnetic Polar
5t6	6	Endcap Tetrahedron
5u10	10	Upper Longitudinal
6m2	2	Magnetic Polar
6t6	6	Endcap Tetrahedron
6u0:10	-0-	Upper Longitudinal
6v14:14	14	Very Big
6e2:3	2	Equatorial

079-Au Gold

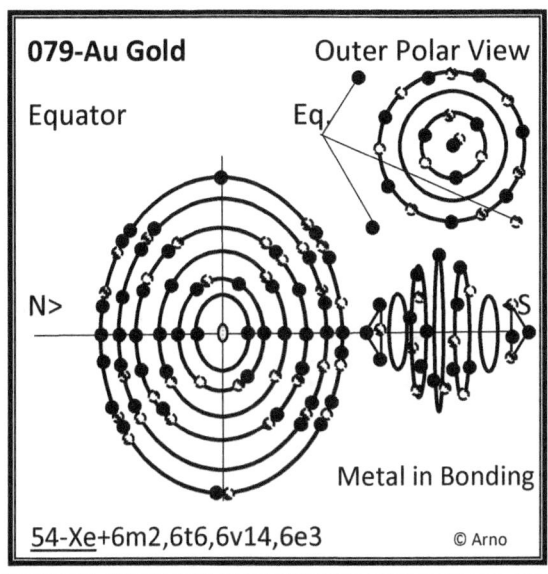

54-Xe+6m2,6t6,6v14,6e3

Shell/Subshell/Count	# Electrons	Structure
1m2	2	Magnetic Polar
2m2	2	Magnetic Polar
2c6	6	Scrunched Cube
3m2	2	Magnetic Polar
3f6	6	Faces of Cube
4m2	2	Magnetic Polar
4t6	6	Endcap Tetrahedron
4u10	10	Upper Longitudinal
5m2	2	Magnetic Polar
5t6	6	Endcap Tetrahedron
5u10	10	Upper Longitudinal
6m2	2	Magnetic Polar
6t6	6	Endcap Tetrahedron
6u0:10	-0-	Upper Longitudinal
6v14	14	Very Big Inclination
6e3:3	3	Equatorial

080-Hg Mercury

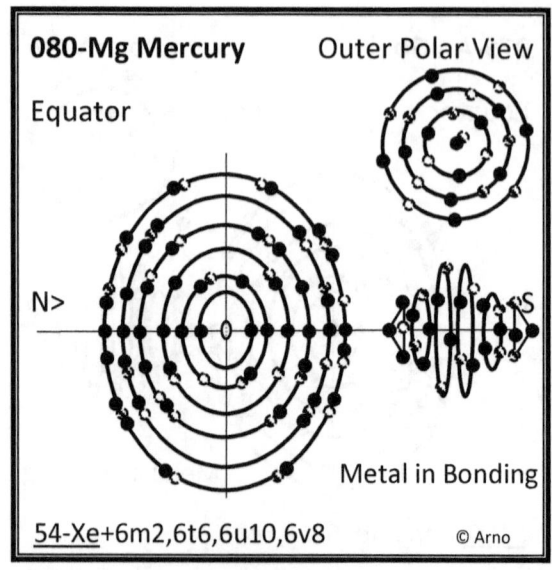

Shell/Subshell/Count	# Electrons	Structure
1m2	2	Magnetic Polar
2m2	2	Magnetic Polar
2c6	6	Scrunched Cube
3m2	2	Magnetic Polar
3f6	6	Faces of Cube
4m2	2	Magnetic Polar
4t6	6	Endcap Tetrahedron
4u10	10	Upper Longitudinal
5m2	2	Magnetic Polar
5t6	6	Endcap Tetrahedron
5u10	10	Upper Longitudinal
6m2	2	Magnetic Polar
6t6	6	Endcap Tetrahedron
6u10	10	Upper Longitudinal
6v8	8	Very Big

081-Tl Thallium

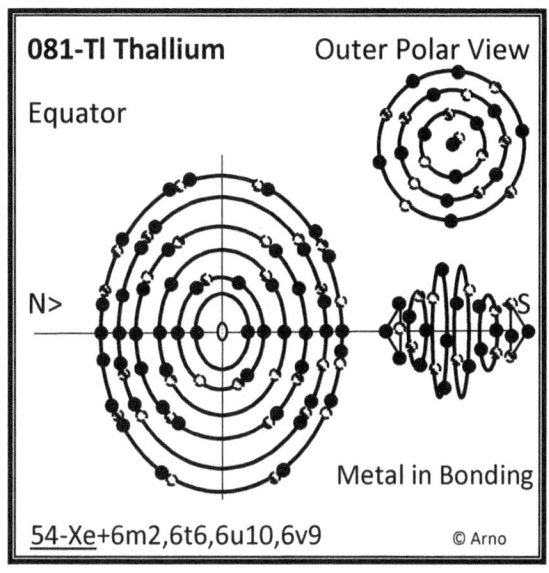

Shell/Subshell/Count	# Electrons	Structure
1m2	2	Magnetic Polar
2m2	2	Magnetic Polar
2c6	6	Scrunched Cube
3m2	2	Magnetic Polar
3f6	6	Faces of Cube
4m2	2	Magnetic Polar
4t6	6	Endcap Tetrahedron
4u10	10	Upper Longitudinal
5m2	2	Magnetic Polar
5t6	6	Endcap Tetrahedron
5u10	10	Upper Longitudinal
6m2	2	Magnetic Polar
6t6	6	Endcap Tetrahedron
6u10	10	Upper Longitudinal
6v9	9	Very Big

082-Pb Lead

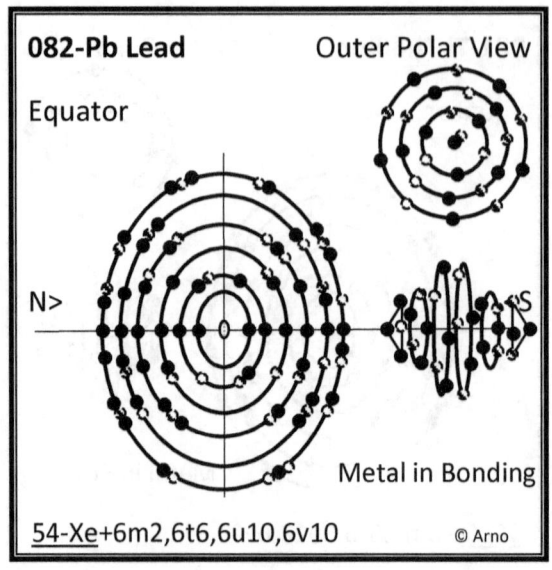

Shell/Subshell/Count	# Electrons	Structure
1m2	2	Magnetic Polar
2m2	2	Magnetic Polar
2c6	6	Scrunched Cube
3m2	2	Magnetic Polar
3f6	6	Faces of Cube
4m2	2	Magnetic Polar
4t6	6	Endcap Tetrahedron
4u10	10	Upper Longitudinal
5m2	2	Magnetic Polar
5t6	6	Endcap Tetrahedron
5u10	10	Upper Longitudinal
6m2	2	Magnetic Polar
6t6	6	Endcap Tetrahedron
6u10	10	Upper Longitudinal
6v10	10	Very Big

083-Bi Bismuth

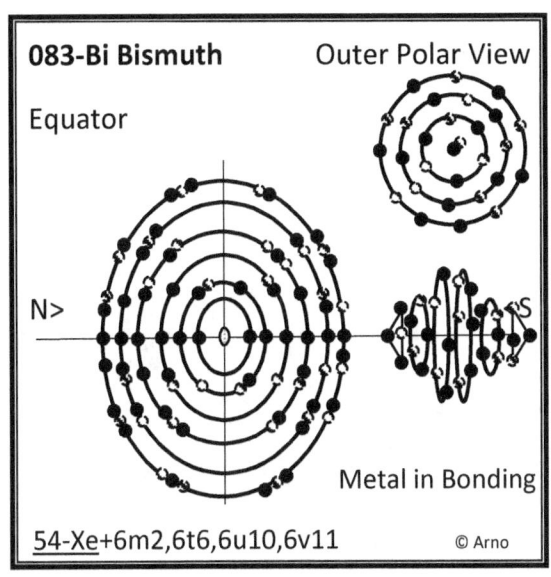

Shell/Subshell/Count	# Electrons	Structure
1m2	2	Magnetic Polar
2m2	2	Magnetic Polar
2c6	6	Scrunched Cube
3m2	2	Magnetic Polar
3f6	6	Faces of Cube
4m2	2	Magnetic Polar
4t6	6	Endcap Tetrahedron
4u10	10	Upper Longitudinal
5m2	2	Magnetic Polar
5t6	6	Endcap Tetrahedron
5u10	10	Upper Longitudinal
6m2	2	Magnetic Polar
6t6	6	Endcap Tetrahedron
6u10	10	Upper Longitudinal
6v11	11	Very Big

084-Po Polonium

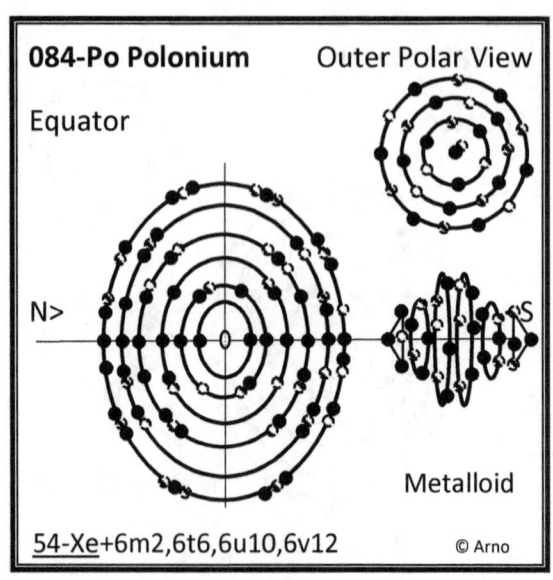

Shell/Subshell/Count	# Electrons	Structure
1m2	2	Magnetic Polar
2m2	2	Magnetic Polar
2c6	6	Scrunched Cube
3m2	2	Magnetic Polar
3f6	6	Faces of Cube
4m2	2	Magnetic Polar
4t6	6	Endcap Tetrahedron
4u10	10	Upper Longitudinal
5m2	2	Magnetic Polar
5t6	6	Endcap Tetrahedron
5u10	10	Upper Longitudinal
6m2	2	Magnetic Polar
6t6	6	Endcap Tetrahedron
6u10	10	Upper Longitudinal
6v12	12	Very Big

085-At Astatine

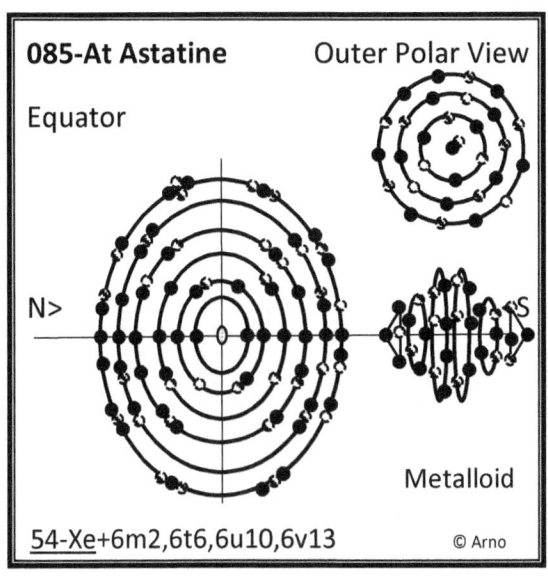

Shell/Subshell/Count	# Electrons	Structure
1m2	2	Magnetic Polar
2m2	2	Magnetic Polar
2c6	6	Scrunched Cube
3m2	2	Magnetic Polar
3f6	6	Faces of Cube
4m2	2	Magnetic Polar
4t6	6	Endcap Tetrahedron
4u10	10	Upper Longitudinal
5m2	2	Magnetic Polar
5t6	6	Endcap Tetrahedron
5u10	10	Upper Longitudinal
6m2	2	Magnetic Polar
6t6	6	Endcap Tetrahedron
6u10	10	Upper Longitudinal
6v13	13	Very Big

086-Rn Radon

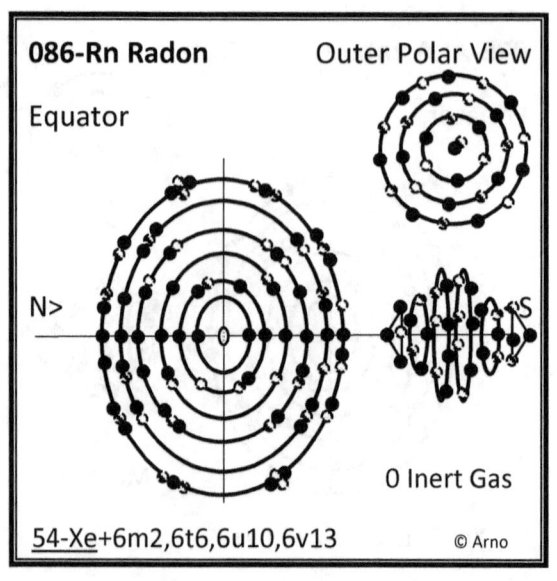

Shell/Subshell/Count	# Electrons	Structure
1m2	2	Magnetic Polar
2m2	2	Magnetic Polar
2c6	6	Scrunched Cube
3m2	2	Magnetic Polar
3f6	6	Faces of Cube
4m2	2	Magnetic Polar
4t6	6	Endcap Tetrahedron
4u10	10	Upper Longitudinal
5m2	2	Magnetic Polar
5t6	6	Endcap Tetrahedron
5u10	10	Upper Longitudinal
6m2	2	Magnetic Polar
6t6	6	Endcap Tetrahedron
6u10	10	Upper Longitudinal
6v14	14	Very Big

087-Fr Francium

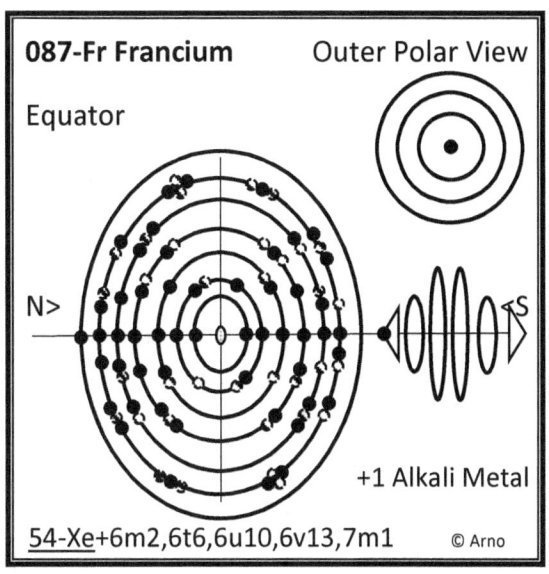

Shell/Subshell/Count	# Electrons	Structure
1m2	2	Magnetic Polar
2m2	2	Magnetic Polar
2c6	6	Scrunched Cube
3m2	2	Magnetic Polar
3f6	6	Faces of Cube
4m2	2	Magnetic Polar
4t6	6	Endcap Tetrahedron
4u10	10	Upper Longitudinal
5m2	2	Magnetic Polar
5t6	6	Endcap Tetrahedron
5u10	10	Upper Longitudinal
6m2	2	Magnetic Polar
6t6	6	Endcap Tetrahedron
6u10	10	Upper Longitudinal
6v14	14	Very Big
7m1	1	Magnetic Polar

088-Ra Radium

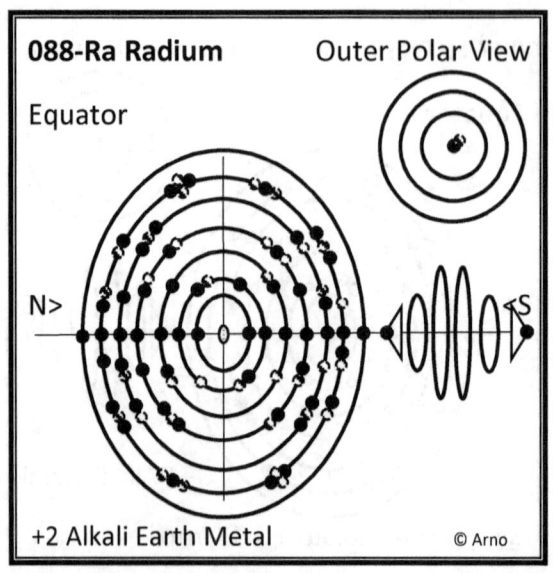

Shell/Subshell/Count	# Electrons	Structure
1m2	2	Magnetic Polar
2m2	2	Magnetic Polar
2c6	6	Scrunched Cube
3m2	2	Magnetic Polar
3f6	6	Faces of Cube
4m2	2	Magnetic Polar
4t6	6	Endcap Tetrahedron
4u10	10	Upper Longitudinal
5m2	2	Magnetic Polar
5t6	6	Endcap Tetrahedron
5u10	10	Upper Longitudinal
6m2	2	Magnetic Polar
6t6	6	Endcap Tetrahedron
6u10	10	Upper Longitudinal
6v14	14	Very Big
7m2	2	Magnetic Polar

089-Ac Actinium

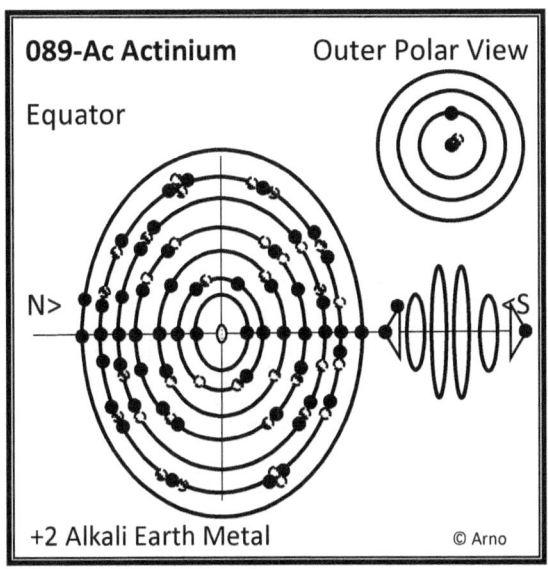

Shell/Subshell/Count	# Electrons	Structure
1m2	2	Magnetic Polar
2m2	2	Magnetic Polar
2c6	6	Scrunched Cube
3m2	2	Magnetic Polar
3f6	6	Faces of Cube
4m2	2	Magnetic Polar
4t6	6	Endcap Tetrahedron
4u10	10	Upper Longitudinal
5m2	2	Magnetic Polar
5t6	6	Endcap Tetrahedron
5u10	10	Upper Longitudinal
6m2	2	Magnetic Polar
6t6	6	Endcap Tetrahedron
6u10	10	Upper Longitudinal
6v14	14	Very Big
7m1	1	Magnetic Polar
7t1	1	Endcap Tetrahedron

090-Th Thorium

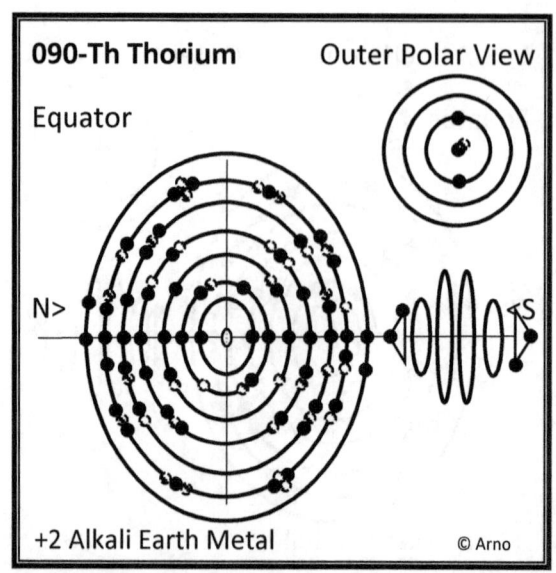

Shell/Subshell/Count	# Electrons	Structure
1m2	2	Magnetic Polar
2m2	2	Magnetic Polar
2c6	6	Scrunched Cube
3m2	2	Magnetic Polar
3f6	6	Faces of Cube
4m2	2	Magnetic Polar
4t6	6	Endcap Tetrahedron
4u10	10	Upper Longitudinal
5m2	2	Magnetic Polar
5t6	6	Endcap Tetrahedron
5u10	10	Upper Longitudinal
6m2	2	Magnetic Polar
6t6	6	Endcap Tetrahedron
6u10	10	Upper Longitudinal
6v14	14	Very Big
7m1	1	Magnetic Polar
7t2	2	Endcap Tetrahedron

091-Pa Protactinium

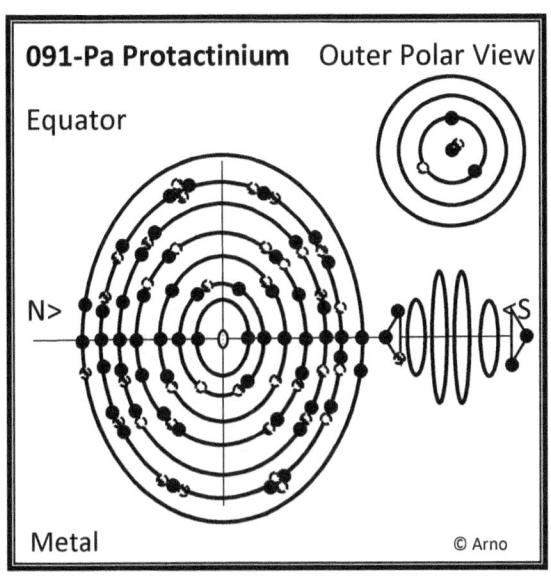

Shell/Subshell/Count	# Electrons	Structure
1m2	2	Magnetic Polar
2m2	2	Magnetic Polar
2c6	6	Scrunched Cube
3m2	2	Magnetic Polar
3f6	6	Faces of Cube
4m2	2	Magnetic Polar
4t6	6	Endcap Tetrahedron
4u10	10	Upper Longitudinal
5m2	2	Magnetic Polar
5t6	6	Endcap Tetrahedron
5u10	10	Upper Longitudinal
6m2	2	Magnetic Polar
6t6	6	Endcap Tetrahedron
6u10	10	Upper Longitudinal
6v14	14	Very Big
7m1	1	Magnetic Polar
7t3	3	Endcap Tetrahedron

092-U Uranium

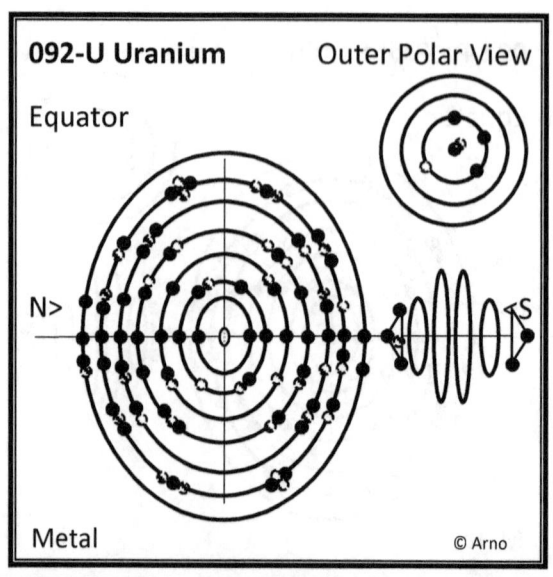

Shell/Subshell/Count	# Electrons	Structure
1m2	2	Magnetic Polar
2m2	2	Magnetic Polar
2c6	6	Scrunched Cube
3m2	2	Magnetic Polar
3f6	6	Faces of Cube
4m2	2	Magnetic Polar
4t6	6	Endcap Tetrahedron
4u10	10	Upper Longitudinal
5m2	2	Magnetic Polar
5t6	6	Endcap Tetrahedron
5u10	10	Upper Longitudinal
6m2	2	Magnetic Polar
6t6	6	Endcap Tetrahedron
6u10	10	Upper Longitudinal
6v14	14	Very Big
7m1	1	Magnetic Polar
7t4	4	Endcap Tetrahedron

Angles by 3D Geometry

For the AVSC model, in a Cartesian coordinate system, we assign the direction based upon the following:

- 'z' = the magnetic poles
- 'x' = this is the base direction. For the first non-polar electron (2e1 or 2c1) x is set at zero (0).
- 'y' = the direction relative the poles of the first non-polar electron (2e1 or 2c1)

In a spherical coordinate system, we assign the direction as follows:

- 'r' and sometimes 'z' if the magnetic poles = the radius from the center of the sphere (the altitude)
- 'i' = The inclination angle from the magnetic poles designated North. That is, at zero (0), this points north or south.
- 'j' = this is longitude relative to the magnetic axis. For the first non-polar electron (2e1 or 2c1), j is set at zero (0) as the atomic meridian.

Electron-Nucleus-Magnetic-Axis by Electron Shell Configuration

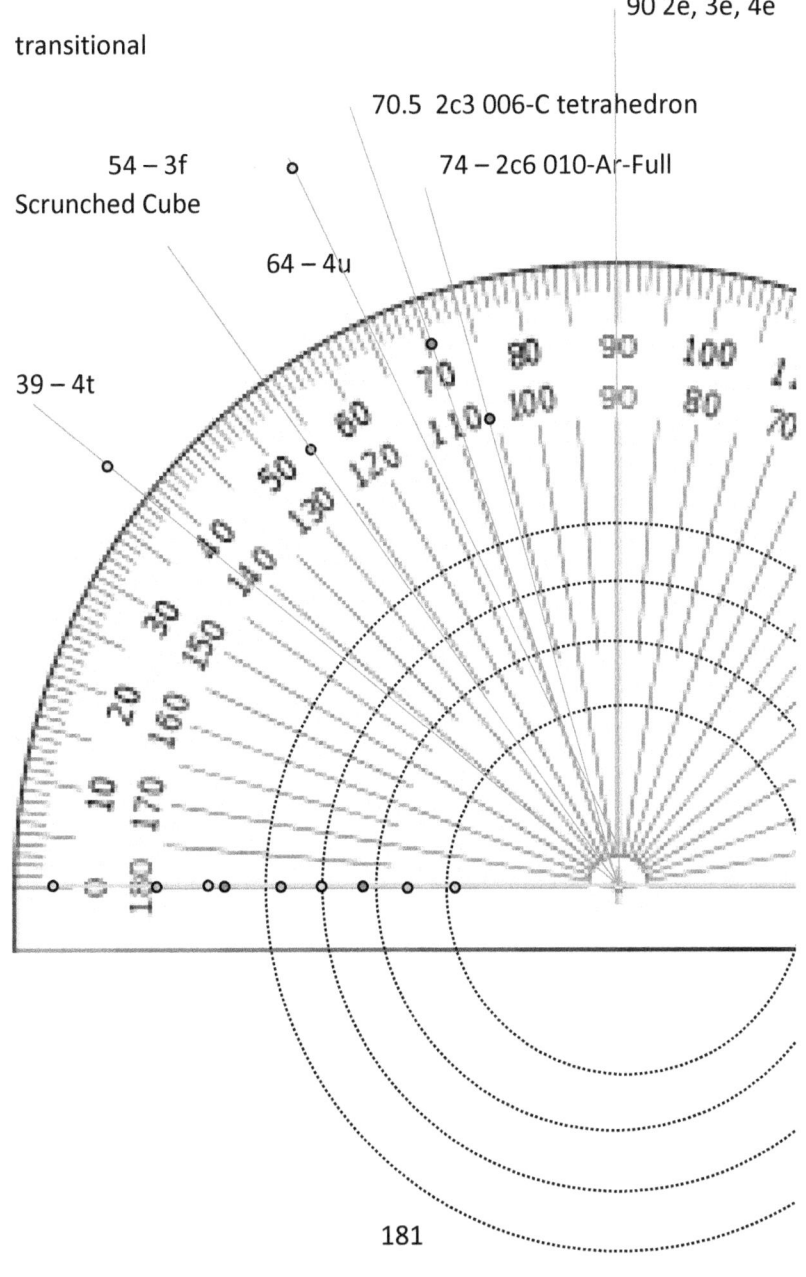

90 2e, 3e, 4e

transitional

70.5 2c3 006-C tetrahedron

54 – 3f
Scrunched Cube

74 – 2c6 010-Ar-Full

64 – 4u

39 – 4t

181

- Full 2-Shell
- Full 3-Shell
- Full 4-Shell

Shell positions (radius-dimension) change as more outer electrons are added.

You can see that as the elements have move nucleus protons, so proton attraction, (net of magnetic repulsion [protons plus neutrons] the radius of each inner shell gets closer and closer.

Subshell 1m

with outer shell   2c   3f   4t,u   Nucleus

Subshell 2m

with outer shell

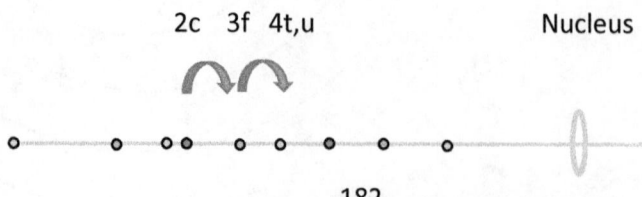

Subshell 3m

with outer shell

        3f 4t,u   Nucleus

Electron-Nucleus-Magnetic-Axis by Electron Shell Configuration

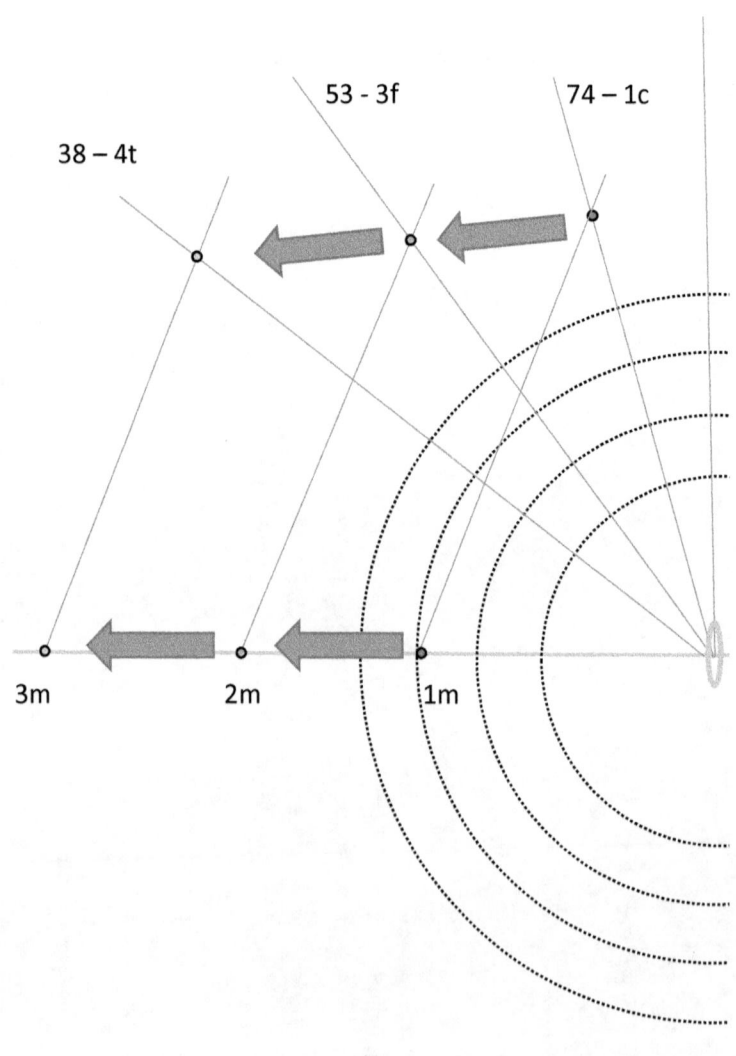

2nd Subshells consistencies

The three (3) electrons in the 2nd subshells is always about 74 degrees off the outer m-subshells electron. That makes them move in generally a line parallel to the magnetic axes. You can see this above. From the e-2c, move in the z-direction (the magnetic axis) (and around 60 degrees in longitude) and get the e-3f electron. Of course, that off 60 degrees means that the electron is really about the same angle versus the m-subshell (not the nucleus). From the e-1c, move in the z-direction (the magnetic axis) (and around 60 degrees in longitude) and get the e-2f electron. Of course, that means the next even electron move 60 degrees which puts it back in line with the e-2c

First, after the m-shells, at each layer, the next subshells (2c, 3f, 4t, 5t, 6t, 7t) all settle at about the same distance off the magnetic axis. Second, those are all 3 electrons in each hemisphere.

The 2nd subshell is always 2 hemispheres by 3 electrons. The full pattern is 1/3/r/s/t/t/s/r/3/1. The 2nd subshell in each Primary Shell basically builds at angles (0,120,24) for even shells, then at offsets (60, 180, 300) for odd-numbered shells – looking at it from each pole – which means they are offset if you look like a full six (6) evenly spaced around the magnetic axis.

For just one subshells in the 2nd position away from the axis:

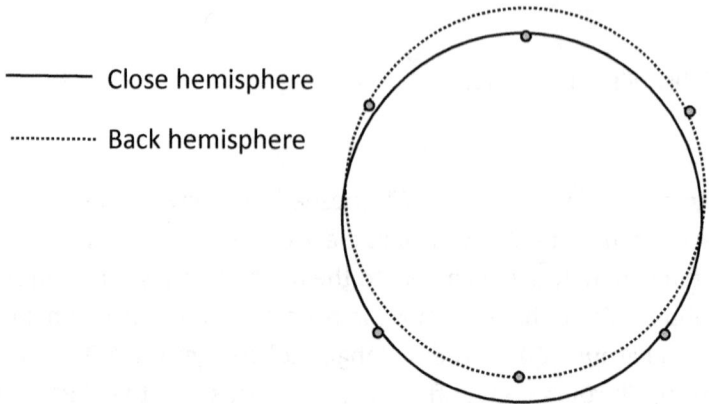

For the building of the subshells in the 2nd position in just one (1) hemisphere, but with 2c, 3f, 4t, and so on the electrons flip flop from 0, 120, 240 to 60, 180, 300, then back to 0, 120, 240:

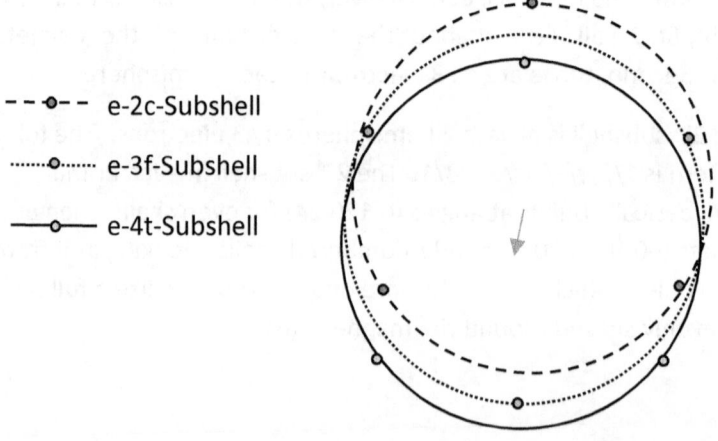

The patterns within shells are also clear. As more particles, the atomic weight, which is mass, gets added, the outer shells get pushed slight further out (at more angle).

Subshell 2c - Corners of the Cube

Element	e-2c-Magnetic Axis Angle
006-C Carbon	70.5
007-N Nitrogen	71.2
008-O Oxygen	72.3
009-Fl Florine	73.3
010-Ar Argon	74.2

006-C 2c1   007-N 2c1

Added particle for 007-N

e-2m2

In each subshells, the location gets pushed out (a greater electron-nucleus-magnetic axis angle) by added electrons. However, the added nucleus particles actually bring the electron slightly closer to the nucleus. These two forces create a) higher angles, and b) shorter distances as Atomic Number and Atomic Weight increase:

        Attraction force for 006-C   6 protons

        Attraction force for 007-N   7 protons

offset by $[½^{(2/3)}]*(1+3COS^2)^{(1/2)}$ because magnetics also increases

        Attraction force for 006-C   12 protons

        Attraction force for 007-N   15 protons

Faces of the cube

At the angle of the 2c1, the 2c2 and 2c3 are half way down, but they sit at an angle (+60 and -60) from the 2c1. Further, the 2f1 will actually sit in that same orientation (x=0). In that way, the angle of the 3f is ½ of the location of the underlying layer (the full 2c6 of 010-Ar).

However, these are based upon an inner shell which already has established angles built on the 010-Ar Argon structure. That means that the angle does not vary as much on odd shells.

Element	e-2c-Magnetic Axis Angle	
006-C Carbon 106 of 010-Ar Argons e-2c6 electrons	53	which is ½ of 74 /
007-N Nitrogen	53	
008-O Oxygen	53	
009-Fl Florine	53	
010-Ar Argon	53	

AVSC and Quantum Mechanics

The prior art describes these in four factors in quantum mechanics:

Quantum Mechanics Description	AVSC Description
Primary Quantum Number (N)	The Shell Number (1, 2, 3, 4, 5, 6, 7)
Azimuth Quantum Number (ℓ)	The number and orientation of electrons in a subshell which describes a) number of electrons in that subshell, and b) the placements of those relative to electrons of other shells, and c) how many electron pairs have filled that subshell. The AVSC also gives the <i (or '<ℓ') longitude angle relative to the magnetic pole from the nucleus, better known as the inclination 'i' longitude and the 'j' latitude (or azimuth looking in the opposite point of view).
External Magnetic Quantum Number (m)	This is 'external' so a knowledgeable person addition to the base AVSC model formula and re-calculate how external magnetics change the bonding energy, distances, angles and such.
Spin Quantum Number (s)	Whether any electron has a paired electron in the opposite hemisphere which is whether both 'j' and j+180 are both populated.

Shell Number

In both the AVSC model and quantum mechanics, the first factor is the same. Yet, the AVSC model has specific reasoning, hemispheres plus x-y-dimensional growth, or x-squared for 2xN-square, for the sequence where quantum mechanics uses the steps as a given without a physical model.

Subshell Angle and Structure

In both the AVSC model and quantum mechanics, the second factor is an angle. However, quantum mechanics uses the name, azimuth, of an angle from the perspective of the viewer, the electron.

Definition of Azimuth

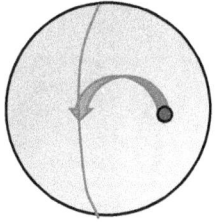

Azimuth is the correct name for what quantum mechanics describes. It is the measured of the angle to the natural axis line from the perspective of the observation, the electron in this case.

The challenge is that azimuth as described in quantum mechanics does not identify either the magnetic axis or the longitudinal prime meridian.

Think of this as figure out the world looking at the west, but you don't know if you are on top of a mountain or in a valley. You don't know if you are in Norway or the Sahara desert. The frame of reference is missing. The words they chose accurately describes their orientation dilemma.

Further, the quantum number is not an angle, but a number. In that sense, quantum mechanics is combining the angle with the structure and also with the count of the subshell in that configuration. It is more the 'spherical drum with XX nodes' in various configuration. As a number, it works very well to describe those results.

AVSC generates the two angles relative to the main axis, the nucleus magnetic axis. The first one is the inclination, '<i' which is the angle relative to the nucleus magnetic axis. The second one is the latitude, '<j' which is the angle relative to the <i inclination of the first non-axis electron (2c1 or 2e1).

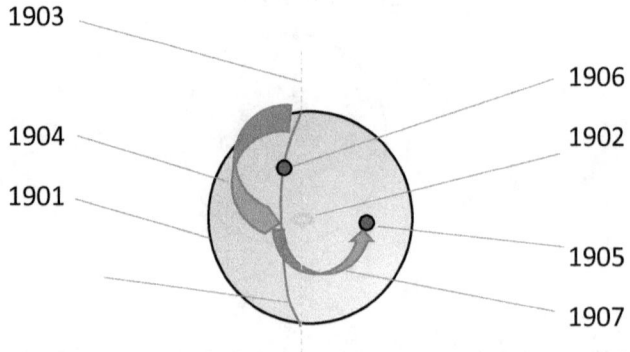

In an atom (1901), with a nucleus (1902) magnetic axis (1903), the angle (1904) from that magnetic pole to the longitude of the chosen electron (1905) is called inclination (<i), and given the first non-polar electron, say e-2c1, (1906) as the prime meridian, the

angle (1907) from that meridian to the chosen electron position is called the longitude (<j).

These positions are defined for each known subshell structure with known adjustments for the counts of electrons within each

Therefore, the subshell name has all of information as the 2nd quantum number, the azimuth quantum number plus additional information:

Attribute	Quantum Mechanics Definition	Defined in ASCV
Longitude	Not defined	<i
Latitude	Not defined	<j
Structure	Harmonics of these are solved in the matrix math solution	Subshell name: 2c, 3m, 4t, etc.
Number of Bodies	Defined by +- (N-1)	Number of electrons in that shell: 2c6, 3m2, 4t6

External Magnetic Quantum Number

This is an additional calculation applied after AVSC. It is not defined specifically because it is external to the AVSC structure set.

Spin Quantum Number

In AVSC, the spin is an attribute of the subshell count (generally about subshells have a count even or odd) to show whether the whole structure has a balance around the magnetic axis. Evens are stable with each electron in one hemisphere with a paired electron in the other hemisphere at 180 degrees across through the nucleus.

In that sense, the quantum 'spin' name is good for stability. However, the actual spin is not the particle, but the entire system's stability.

Quantum mechanics through whole numbers generates certain information well. AVSC can get integrated into that, yet it also provides additional information to address attributes and solutions that quantum mechanics does not address.

Endnotes

[i] *Niels Bohr (1913). "On the Constitution of Atoms and Molecules, Part II Systems Containing Only a Single Nucleus" (PDF). Philosophical Magazine. **26** (153): 476–502. doi:10.1080/14786441308634993.*

www.ingramcontent.com/pod-product-compliance
Lightning Source LLC
Chambersburg PA
CBHW071425180526
45170CB00001B/229